CLOCKMAKERS

OF LANCASTER COUNTY AND THEIR CLOCKS 1750-1850

Lancaster's second courthouse, home of John Eberman, Jr. tower clock. Camera obscura watercolor by W. Banton about 1812 to 1815. (Lancaster County Historical Society)

CLOCKMAKERS

OF LANCASTER COUNTY AND THEIR CLOCKS 1750-1850

Stacy B. C. Wood, Jr. / Stephen E. Kramer III

with a Study of Lancaster County Clock Cases by John J. Snyder, Jr.

VNR VAN NOSTRAND REINHOLD COMPANY
New York Cincinnati Toronto London Melbourne

Dedication

To the clockmakers of Lancaster County, Pennsylvania, from Atkinson to Zahm, who, through their pride of workmanship and sheer numbers, made their county a clock center of the nation for one hundred years.

Library of Congress Catalog Card Number 76-30279
ISBN 0-442-29531-6

Printed in the United States of America.
Designed by Loudan Enterprises

Published in 1977 by Van Nostrand Reinhold Company
A division of Litton Educational Publishing, Inc.
450 West 33rd Street, New York, NY 10001, U.S.A.

Van Nostrand Reinhold Limited
1410 Birchmount Road, Scarborough, Ontario M1P 2E7
Canada

Van Nostrand Reinhold Australia Pty. Limited
17 Queen Street, Mitcham, Victoria 3132, Australia

Van Nostrand Reinhold Company Limited
Molly Millars Lane, Wokingham, Berkshire, England

16 15 14 13 12 11 10 9 8 7 6 5 4 3 2 1

Library of Congress Cataloging in Publication Data

Wood, Stacy B C
Clockmakers of Lancaster County and their clocks, 1750–1850.

Includes bibliographical references and index.
1. Clock and watch makers—Lancaster Co., Pa.
2. Clock and watchmaking—Lancaster Co., Pa.—History.
3. Lancaster Co., Pa.—Biography I. Kramer, Stephen E. joint author. II. Snyder, John J., 1946–
III. Title.
TS543.U6W66 1977 681′.113′ 0922 [B] 76-30279
ISBN 0-442-29531-6

CONTENTS

ACKNOWLEDGMENTS

Help came in many forms from many people during this project. Paramount was that given by the Lancaster County Historical Society, not only through the treasure trove it has in early tax records, newspapers, manuscripts, estate inventories, etc., but also through its cooperation that made access possible to many of the clocks included herein. We also thank the Society's staff, especially their librarian, Mrs. Charles Lundgren.

Thanks, too, to: the Lancaster County offices of the Recorder of Deeds and Register of Wills; Pennsylvania Farm Museum of Landis Valley; Pennsylvania Historical and Museum Commission, Ephrata Cloister; James Buchanan Foundation for the Preservation of Wheatland; Lancaster Mennonite Conference Historical Society; Colonial Donegal Mills Plantation; Heritage Center of Lancaster County, Inc.; Hershey Museum of American Life; Mill Bridge Craft Village/Lefever Historical Society; Amish Farm and House; Earl T. Strickler Collection; Historical Society of Pennsylvania; Historical Society of York County; Historical Society of Berks County; Frick Art Reference Library; Virginia Museum of Fine Arts; Henry Francis du Pont Winterthur Museum; and the National Association of Watch and Clock Collectors.

Finally, thanks to: Samuel E. Dyke; Gerald S. Lestz of *Lancaster New Era* and the Lancaster Newspapers, Inc.; Richard W. Case of Tarrytown, N.Y.; fellow horologists Richard Thomson and Chris Bailey for their encouragement, and James W. Gibbs, Henry B. Fried, and Britain's Brian Loomes for their willingness to share information; the descendants of Abrum Laroy, Samuel Breneisen, Thomas B. Burrowes, A. W. Carpenter, John Eberman, John Esterle, George Ford, Christian Forrer, and George Hoff who provided genealogical information and photographs of their forebears; and the numerous collectors who allowed their clocks to be studied and photographed.

PREFACE

One might ask, "Why a book about clockmakers of so small a geographical area as a county?" Such projects in the United States have historically covered the clocks of the nation or, at least, of a state. A notable exception is *Chester County Clocks and Their Makers* (1947), by Arthur E. James. When I settled in Lancaster County, Pennsylvania, after graduating from the clockmaker's course at Lancaster's ninety-year-old Bowman Technical School, I was astounded to discover that there were nearly one hundred clockmakers working within the present boundaries of the county between 1750 and 1850, a period when clocks were still handcrafted. I also discovered that many examples of their work are to be found throughout the United States, both in private homes and in some of the great museums.

The last major research on this subject was done sixty years ago, and then only as a paper read before the Lancaster County Historical Society. The paper, *Grandfather Clocks: Their Making and Their Makers in Lancaster County* by D. F. Magee (Lancaster County Historical Society Papers XXI, No. 4, 1917), was subsequently enlarged to twenty-seven pages and reread in 1939 (Lancaster County Historical Society Papers XLIII, No. 5, 1939). This paper became the major source for information about clockmakers of Lancaster County after that date.

While addressing a local chapter of the National Association of Watch and Clock Collectors a few years ago, I asked, "Who will take on the work of doing for Lancaster County what Doctor James has done for Chester County?" The late horologist of world renown, Earl T. Strickler, replied, "Why don't you?" I filed that suggestion away with other things that I might do in my retirement some twenty years hence, if the project still remained undone.

In early summer of 1975 it occurred to me that a "driving clock tour" of Lancaster County might be of interest. The idea was to record by word and film those clocks that are on public display in the county. Many fine examples of English, French, Dutch, New England, and Philadelphia clocks, in addition to clocks of Lancaster County, can be seen in the various town historical societies, banks, museums, retirement homes, funeral homes, etc. There are also a number of tower clocks to be found. The most notable of these, after the 1785 courthouse clock, is perhaps the Gap Town Clock which, for the past hundred years, has survived threats of a sheriff's sale, fire, and a highway relocation through its very foundation. I approached fellow Bowman graduate Stephen E. Kramer, III who had accumulated some photographic equipment with the thought of making a detailed study of Philadelphia's famous clockmaker, astronomer, and patriot, David Rittenhouse. (Steve has recently received a grant from the American Philosophical Society to pursue that project.) Within a week, we had obtained the blessing of the Lancaster County Historical Society and started off on our tour.

Some six months later, it became obvious to us that the number of Lancaster County clockmakers represented was expanding to the point that, if private collections were included, we might well have ourselves that book on Lancaster County clockmakers. The decision was made, and we shelved the "foreign" clocks for the time being, and once again with the Historical Society's blessing, started on the trail of county clockmakers and their clocks. Without that blessing, I might add, many doors would not have been opened to us. I soon found myself wishing that I could take up a schizophrenic residence in the archive rooms of the county courthouse and the Historical Society.

Steve, in the meantime, had moved to Philadelphia, but we nevertheless hit the clock trail on a weekly basis. Upon completion of the project (which really translates as "after all leads had been exhausted"), Steve had photographed ninety-seven clocks representing forty-seven of the ninety-four craftsmen that I could identify as clockmakers through tax records, deeds, wills, old newspapers, church records, manuscripts, family genealogies, etc. Unless otherwise identified, all photographs were taken and processed by Steve.

To properly handle the section on clock cases, fellow Lancastrian John J. Snyder, Jr., a recognized expert on southeast Pennsylvania cabinetmakers and their work, was invited to join in. John is a Winterthur Fellow and has written articles about Lancaster County Chippendale furniture in *Antiques* magazine. We are grateful not only for his written contribution, but also for his help in locating some of the more exquisite examples of casework.

The clocks made in Lancaster County really do not differ greatly mechanically from those that were made immediately following the application of Galileo's pendulum theory to clocks ca. 1656, nor do they differ much from the mechanical clocks that are made today. Many readers may have their first look at what lies behind a handsome dial. There may even be a surprise for the sophisticated horologist. In any event, we hope that this book will bring due recognition to this early clock center of America, Lancaster County, Pennsylvania.

INTRODUCTION

To those whose knowledge of Lancaster County is limited to the timeless scenes of bucolic splendor, where rich and rolling limestone farmlands bear weighty harvests at the hands of the Amish Mennonite farmers, the intrusion of the mechanical mind geared directly to insatiated intellectual curiosity seems incredible. Despite such incongruity the truth is that "tinkering with things" has been a characteristic of Lancaster County from its earliest days. That "necessity is the mother of invention" is axiomatic; local mechanics, farmers, and tradesmen, living and working on America's eighteenth century frontier, had countless needs begging invention.

The authors of this book have taken for their survey timekeeping devices extant in Lancaster County. While this field of the mechanical and scientific art may appear limited, nowhere does the state of the art permit pulse-taking with the accuracy to be found in horological invention, design, and production. Moreover, it tells us much about the persons who bought, used, and preserved these timekeepers. With some unorthodox perception it can be argued this is a seminal work, and that it ought to blaze a trail for other works on mechanical devices and systems.

Portrayed in this work are mechanisms produced here. Of importance to us is the need expressed by Lancaster Countians and their institutions in their acquisition of the timekeepers, and their preferences for certain choices available in the mechanical arts. The authors, with their respective talents in horology and photography, have presented, indeed, a social history through the medium of clocks. That they have been successful in accomplishing that objective and others with ingenuity, style, discernment, and visual appeal is a tribute to their zeal and sense of history.

In the history of mankind a sense of order and regularity had to become established before society could organize itself for protection and nourishment. Certainly one of the earliest known uses of timekeepers was in communal living such as in monasteries. The times for work, prayer, nourishment, worship, and private reflection had to be set, and that necessitated some means of telling time. Only by the cultivation of time habits—scheduling—could early man be made productive and useful, and his ways subjected to the discipline of regularity. Later, during the development of industrialized society, efficiency became a factor in human productive functions. And again the clock was the center of this efficiency. It would not be stretching a point to say the timekeeping mechanism was every bit as essential as the steam engine in creating our industrial society.

Clocks and watches tell us more than time's passage; we are impressed with the independence of the timepiece. The clocks we know usually present time for our interpretation in the form of symbols. Twelve hours, divided into minutes and seconds, are the most common. But other timepieces tell time using other symbols. Ship watches are not understood by most laymen, and electronic and scientific timekeeping devices abound in our world today. Regardless of the language "spoken" by the clock, it performs the same function: marking the passage of time, accurately, impersonally, amorally, with perfect passionless cadence. Indeed, the clock itself, created originally to serve man, became a model, and later, a master of humanity. A sense of time did not dwell among the classical peoples of Greece and Rome, according to Bury and other historians of those ages. With the invention of the clock, man reportedly began dividing his past into periods. Possessed with a new-found time consciousness, man saw his ancestors' achievements happening during specific eras rather than as summits in a continuum of time.

A mechanical clock contains examples of every simple machine; its trains of gears perform marvelously with predetermined mathematical precision, emulating only the planetary movement untouched by the hand of man in this universe. Man—that creature that possesses in his dualistic nature the rational as well as the irrational—admires and tries to imitate the perfect order

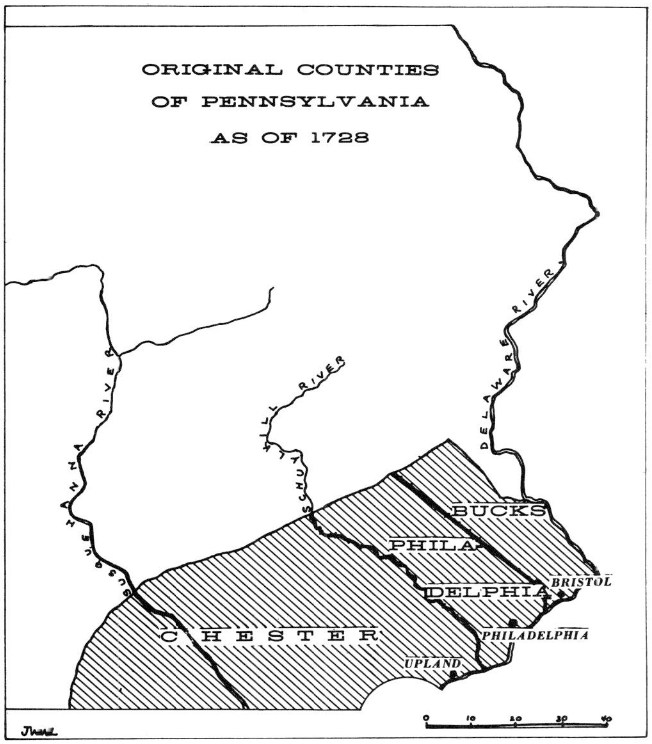

Original counties of Pennsylvania just before formation of Lancaster County in 1729. (Lancaster County Historical Society, Journal Research Staff)

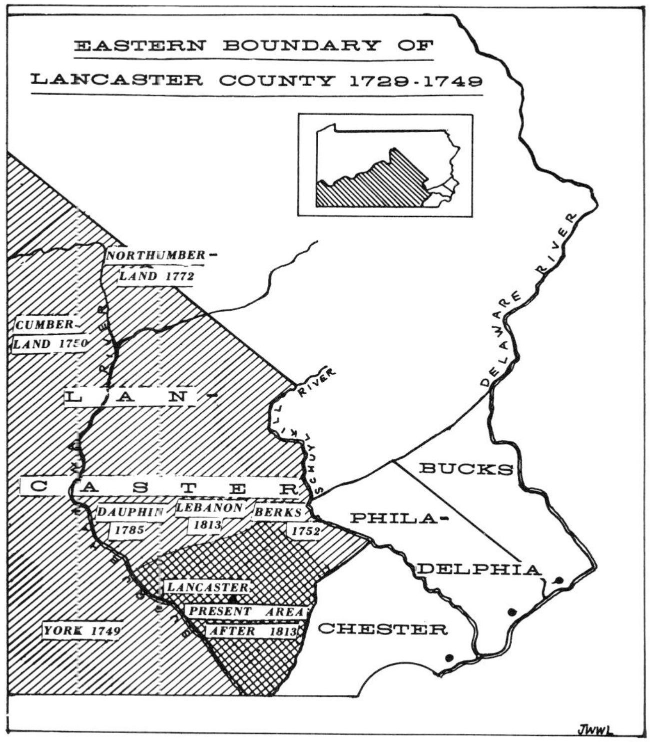

Eastern portion of Lancaster County as it appeared from 1729 to 1749. Entire extent of Lancaster County during that period is shown in insert map. (Lancaster County Historical Society, Journal Research Staff)

found in nature. To him the chronometric instrument is a thing of awe, and perhaps, an idol worthy of worship. It is the nearest thing visible to him that reproduces the natural order. Timepieces take on personalities, it would seem, which may in some mystical way be associated with our lives, or the lives we would like to live. They stand like ticking critics, reminding us of the precious moments we waste, and tempering our pride when we have used our time wisely. From the labor room of the maternity ward to the mortuary, our lives are unwitting companions, even prisoners, of timepieces.

Those clocks that strike time periods have an additional dimension to their personalities, as anyone living in a community with a tower clock well knows. The clock in the tower of the ill-fated Columbia Opera House marked the hours by intoning a deep knell not unlike what an imaginative youngster would attribute to the sound of the gates of perdition clanging shut. A quaint, insistent tenor strikes the quarter, half, and three-quarter hours from the tower of historic Lititz Moravian Church. Some bells seem happy, others sad, some brassy and loud, others prim and sedate. How many times have communities and churches been judged by the sounds of their clock bells?

The Lancaster Countian, that rugged individualist that defies all attempts at classification, always has been a practical and selfsufficient person. He is proud of his ability to make what he needs, and contrive something new when known devices are inadequate. Local farmers and artisans have been interested keenly in the mechanical arts since 1710 when the first permanent settlement was made by Swiss Mennonites in what became Lancaster County. Within a decade the entire area between the Octoraro Creek and the Susquehanna River and along the valleys of the Conestoga River and Pequea Creek were peopled by Scot Presbyterians from Ulster, French Huguenots, English and Welsh Quakers and Anglicans, and Germans from the various principalities, kingdoms, and states of the Rhineland and Bavaria. Among the Germans were Lutherans, Calvinists (or "Reformed"), and Anabaptists that became either Mennonites or Baptist Brethren. Followers of Count Zinzendorf established Moravian communities in eastern Pennsylvania, including the famed settlement of Lititz. By 1750 Lancaster County had become the most pluralistic place in the Western Hemisphere. Roman Catholics and Jews completed the mosaic of ethnic and religious diversity. To an extraordinary degree all these people practiced tolerance toward each other, their energies being required wholly for the common defense against the frontier wilderness surrounding them.

This was no place for the lazy and unskilled. Every man had to be a farmer and an artisan of sorts. What was not available he made himself. By the 1750s and the French and Indian Wars, Lancaster County had become the military manufacturing and supply depot of the middle colonies. With the coming of the American Revolution Lancaster artisans were experienced in the manufacture of rifles and muskets, iron hardware, leather goods, wagons, clothing, glass, furniture, and clocks. William Henry, a Moravian riflemaker, also invented a screw auger and tinkered with paddlewheel boats in the 1770s. Robert Fulton, born in Lancaster County, was a veritable fountain of inventive genius. County farmers invented farming machinery including a steam plow. With the coming of railroads more persons threw themselves into developing all manner of safety devices. One of Lancaster's early large industries was the manufacture of steam locomotives.

Even colonial Lancaster had its numerous instrumentmakers. Its skilled cabinetmakers—whose elegant furniture rivaled the best designed and made in Philadelphia, New York, and Boston—teamed up with clockmakers to produce those handsome and magnificent tall clocks about which this volume speaks so eloquently. The watchmaking industry, now over a century old in Lancaster, has gained a worldwide reputation. Certainly it was no accident that the renowned Bowman Technical School should be located here.

The Lancaster County Mind loves order—order in the lives of the people, and order in mechanical devices. Nothing could be more appropriate than to perceive the insights of this county and its people through the timepieces they have created and bought, and have used and appreciated these many years. Time flies, but when over Lancaster County it dips a wing in tribute. A tribute is in order to Stacy B. C. Wood, Jr., and Stephen E. Kramer, III for a work of great merit done with unabashed love and dedication.

John Ward Willson Loose
President, Lancaster County Historical Society

I DIRECTORY OF CLOCKMAKERS

The following directory contains the names of ninety-four clockmakers who conducted business between 1750 and 1850 within the present boundaries of Lancaster County, Pennsylvania. This date span encompasses the period in which clocks were generally handcrafted. The boundaries were chosen because earlier Lancaster County extended as far west as Ohio and included such present-day counties as Berks, Lebanon, Cumberland, and York.

Each clockmaker's name is given as it has appeared on the dial of one of his clocks or in advertisements in the English language newspapers of the day. Although many of the early clockmakers came from countries other than England, the spelling is anglicized, e.g., Forrer instead of Fuhrer and Shreiner rather than Schreiner.

County tax records were a primary source in substantiating the trade of clockmaker. Unfortunately, the records of some years, especially after 1820, have not survived and there are gaps within other decades. Some townships listed trades while others did not. Registered deeds were also of help for the early ones often specified the trades of the parties involved. Wills provided similar information, often specifying to whom the tools were to be left. Inventories of the estates of the deceased naturally provided the greatest evidence. Newspaper ads and other surviving manuscripts also gave factual evidence.

Discrepancies often exist between the dates of arrival in America of the earliest clockmakers and their first appearance in county tax records. This cannot be blamed on fragmented records alone, for often the immigrant was forced into an indenture in order to pay for his transatlantic voyage. If he were fortunate, he would work in his trade. A check of Philadelphia tax records shows that the clockmaker was not temporarily in business there, conversely, if he were indentured he would not be taxed and therefore would not appear in the tax records.

Although these were indeed clockmakers, many were also tavern- and innkeepers and, therefore, they often moved from town to town, for being an innkeeper did not necessarily mean that one owned the inn: more often he was in a managerial position. Apparently the work load was such that an innkeeper had time to have a secondary trade such as clockmaker. Over ten percent of the following clockmakers were in fact innkeepers. With Lancaster being a last frontier town before the wilds, it gave the enterprising innkeeper a chance to make a few sales.

Other clockmakers were involved with precious and semiprecious metals and provided the townspeople with buckles, seals, rings, and jewelry. Most were also watchmakers who imported partially made English watches, during times of good trade relations, and finished them. A few of the earliest clockmakers, such as Rudy Stoner, may have made watches in entirety. Needless to say, these men were clock and watchmakers with their own skills and specialized inventory of tools, both of which were quite different from those of the joiner or cabinetmaker who would be engaged to make the case for the clock.

The clockmakers were often civic minded. They became some of the most highly regarded citizens of the town and on many occasions were elected to town, county, and state offices. Others were active in their churches, volunteer fire companies, free libraries, and school systems.

One did not become a clockmaker overnight. Usually one became apprenticed at the age of fourteen or so to an established clockmaker. The following indenture of John Erb (b. 1814, d. ca. 1860) is an example of what master and apprentice might expect of one another.

"This Indenture Witnesseth; That JOHN ERB aged fifteen year & ten months on the 2nd Day of August 1830 Son of Joseph Erb of Lampeter Township, Lancaster County, & by and with the advise & consent of his father hath put himself and by these presents doth, voluntarily, and of his own free will and accord, put himself Apprentice to JOSEPH BOWMAN of the

LANCASTER COUNTY 1977

Lancaster County, 1977. (Lancaster County Historical Society, Journal Research Staff)

Borough of Strasburg, County of Lancaster, Clockmaker, to learn clockmaking art, trade, and mystery; and after the manner of an Apprentice, to serve five years and two months from the date hereof, for and during and to the full end term, of five years and two months next ensuing. During all which term, the said Apprentice his said Master faithfully shall serve; his secrets keep, his lawful commands every where readily obey. He shall do no damage to his said Master's goods, nor lend them unlawfully to any. He shall not commit fornication, nor contract matrimony, within the said term. He shall not play at cards, dice, or any other unlawful game, whereby his master may have damage. With his own goods, nor the goods of others, without license from his said Master he shall neither buy nor sell. He shall not absent himself day nor night, from his said Master's service without leave; nor haunt alehouses, taverns, or playhouses; but in all things behave himself as a faithful Apprentice ought to do, during the said term, of five years and two months.

"And the said Master shall use the utmost of his endeavors to teach or cause to be taught or instructed, the said Apprentice, in the trade or mystery of Clockmaking . . . and procure and provide for him sufficient meat, drink, lodging, and washing, fitting for any Apprentice, during the said term of five year and two months and said Master shall give said Apprentice within said apprenticeship Six Months Schooling in an English Day School and at the expiration of said term said master shall give said Apprentice twenty-five Dollars Good and Lawful Money Which is to be in Lue of his freedom suit.

"And, for the true performance of all and singular the covenants and agreements aforesaid, the said Parties bind themselves each unto the other, firmly by these presents.

"IN WITNESS whereof, the said Parties have, interchangeably set their hands and seals hereunto. Dated the twenty-sixth day of June in the year of our Lord, one thousand eight hundred and thirty.

Sealed and delivered in the presence of

William Black	John Erb (Seal)
John Markley	Joseph Erb (Seal)
	Joseph Bowman (Seal)"

It is unfortunate that this is the only known Lancaster County indenture of a clockmaker extant. Even the original copy's whereabouts is now unknown. Following satisfaction of the indenture, the new clockmaker might initially hire on as a journeyman to yet another clockmaker, not necessarily, however, one who had been in business very long. If he could find sufficient funds, he might go into business for himself. The rule of thumb for Lancaster County clockmakers, however, was that they started their own businesses in their twenty-third year. This is indicated by tax records and newspaper ads. One must realize that for many of them their lives were already two-thirds over. Quite often, they would marry and start families shortly after starting into business.

Some names that have appeared in previously published listings have been omitted from the following directory because they could not be substantiated as clockmakers. Approximately fifty percent of those who are listed are represented by at least one clock movement in Part IV. Technical aspects of their movement design will generally be found in that section.

Atkinson, Wilton. *Lancaster.*

Believed to be a son of Matthew Atkinson of Lampeter Township. Various manuscripts, ledgers, and deeds indicate that Wilton Atkinson was in the Borough as early as 1763 as a cutler. He appears in the tax records as a watchmaker in 1770 and as a clockmaker in 1772 and 1773. Atkinson married Anna Maria, a daughter of the early Borough clockmaker Abrum Laroy and legend states that she also made clocks in Lancaster that were sold bearing her husband's name. They eventually moved to Baltimore, Maryland where, following his death, Anna Maria conducted business until 1819.

Ayars, Hamilton. *New Holland.* (b. 1816, d. ca. 1860)

Appears in the voting list of 1841 and the census of 1850. The Lancaster City Directory of 1863–1864 lists his widow as residing on North Prince Street below Walnut. Only a few of his clocks survive.

Baldwin, Anthony Wayne. *Lampeter Square.* (b. 1783, d. 1867)

Baldwin was born in Chester County, a son of sometime silversmith Anthony Baldwin and his wife Mary Harlan. He was a brother of George B. Baldwin (1777–1844), a clockmaker in Sadsburyville, Chester County. Anthony Wayne Baldwin learned the trade from his father-in-law, Joseph Bowman, Sr., and in turn taught his brother-in-law Joseph Bowman, Jr. He first appears in the Lampeter Township tax records of 1815 although he is known to have lived there as early as 1810. A son, John C., born in Lampeter in 1810, became a watchmaker. Baldwin's dials are signed "A. W. Baldwin."

Barninger, Daniel. (b. 1786)

From 1821 to 1845 his occupation was listed as "clockcleaner" in the Conestoga Township tax records. The list of 1858 records him as a clock- and watchmaker. It is doubtful that Barninger made any clocks.

Belsner, John. *Ephrata.* (d. May 21, 1785)

Appears in the Ephrata Cloister death register as a Brother although he is not believed to have lived within the Cloister.

Bowman, Joseph (Jr.) *Strasburg.* (b. Dec. 10, 1799, d. Apr. 14, 1892)

Bowman was born in New Holland, a son of clockmaker Joseph Bowman, Sr. He moved with his parents to Lampeter Township around 1807. At the age of fourteen he was apprenticed to his brother-in-law, clockmaker A. W. Baldwin. By 1821 he had set up his own business in Strasburg on the south side of West Main Street, next door to the building presently housing the *Strasburg News*. In 1830 he took on John Erb as apprentice. His own son, Edmund (b. 1823), became a watchmaker. Bowman served as borough assistant burgess and burgess off and on for a period of forty-six years. Since he numbered his clocks, it is known that he made at least sixty-seven.

It is said that most of the cases housing his clocks were made by Andrew Charles, although the County Directory of 1869 states that Bowman was also a cabinetmaker.

Bowman, Joseph (Sr.) *New Holland.* (d. 1811)

Bowman taught his son-in-law, Anthony Wayne Baldwin, the art of clockmaking. He moved to Elliot's Corner in Lampeter Township about 1807 and farmed until his death. He was the father of Joseph Bowman, Jr., another clockmaker. No known clocks by Bowman survive.

Boyter, Daniel. *Lancaster.*

Advertised in the *Lancaster Journal* of August 9, 1805 as having come from England and being located "next to the stage-office, East King-street." He offered his services in the repair of "repeating and common watches, clocks, and time-pieces." Boyter does not appear in the Borough tax records and it would appear that his stay in Lancaster was short lived. No clocks bearing his name have been located.

Bradycamp, Lewis. *Lancaster.* (b. May 10, 1815, d. Sept. 5, 1870)

Also known as Louis Bradecamp, this clockmaker came to America from Germany. He was located in Donegal Township in 1850 and had relocated to Lancaster City by 1863, where he had a house on Mulberry Street near Chestnut. He served as a private in Co. F, 50th Regt. and was discharged in February 1862 on a surgeon's certificate. No known clocks survive.

Breneisen, George. *Adamstown.* (b. ca. 1800)

Appears in the Cocalico Township tax records from 1825 through 1845. In 1838 he was in partnership with "Bach" (probably the blacksmith of the same name). He was a son of Samuel Breneisen and a brother of Samuel, Jr. and William who also made clocks.

Breneisen, Samuel. *Adamstown* and *Reamstown.* (b. Feb. 18, 1768, d. Oct. 8, 1837)

Breneisen moved to Cocalico Township in 1810 from Reading where he had worked as early as 1799. He settled in Adamstown where he ran the Rogers Inn. Sometime after 1830 he relocated to Reamstown in the same township where he rented a log house on Main Street. The house and the barn behind it, where he is said to have made his clocks, still stand. Sons Samuel, George, and William also made clocks.

Breneisen, Samuel, Jr. *Adamstown* and *Reamstown*. (b. ca. 1795)

Breneisen accompanied his parents to Adamstown from Reading in 1810. He appears in the Cocalico/East Cocalico tax records for 1825 through 1846 normally with the trade of "tailor." Only in 1828 and 1836 is it indicated that he was a clockmaker like his father, and then only by ditto marks under his father's trade listing. He and his wife and daughter are believed to have moved to Reamstown after 1830. An eight-day tall clock signed "Samuel Breneisen/Reamstown" and having "S B 1834 1" punched into the front plate of the brass movement is thought to have been made by him.

Breneisen, William Schwartz. *Reamstown.* (b. Nov. 14, 1815, d. Oct. 15, 1898)

One of four sons of clockmaker Samuel Breneisen (Sr.), William's trade is listed as a watchmaker between 1842 and 1846, however a clock signed "Samuel Breneisen/Reamstown" with "W B 1834" punched into the front plate of the brass movement is believed to have actually been made by him. The 1860 census shows that William and his wife Sarah (nee Wilson) were living with his mother in Reamstown. After the death of both his wife and mother in the late 1860s, he moved to Gerardstown, West Virginia. He is buried in the Presbyterian cemetery there.

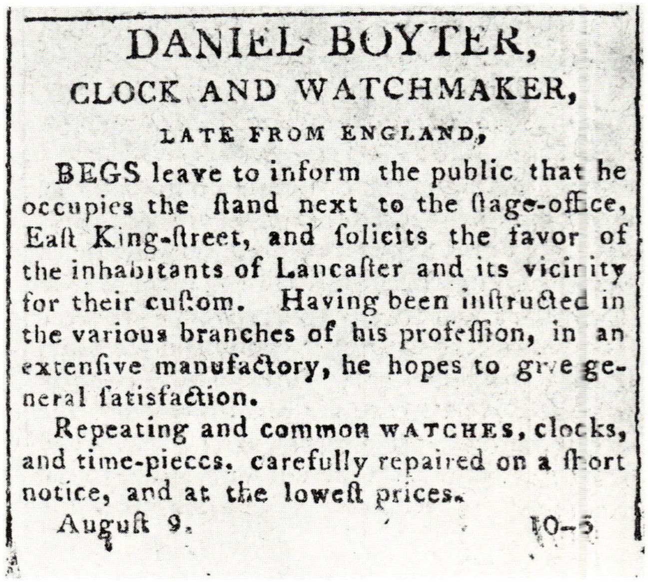

DANIEL BOYTER,

CLOCK AND WATCHMAKER,

LATE FROM ENGLAND,

BEGS leave to inform the public that he occupies the ſtand next to the ſtage-office, Eaſt King-ſtreet, and ſolicits the favor of the inhabitants of Lancaſter and its vicinity for their cuſtom. Having been inſtructed in the various branches of his profeſſion, in an extenſive manufactory, he hopes to give general ſatisfaction.

Repeating and common WATCHES, clocks, and time-pieces, carefully repaired on a ſhort notice, and at the loweſt prices.

Auguſt 9. 10–5.

1-1. Daniel Boyter ad, *Lancaster Journal*, Aug. 9, 1805.

Brown, John. *Lancaster*. (b, 1796, d. ca. 1860)

First noted in the tax records of 1824. His first shop was located at the corner of East King and Christian Streets. By 1840 he had moved to North Queen Street. He also owned a stagecoach office and a tavern.

Burg, Jacob. *Lancaster*. (b. ca. 1790, d. Aug. 3, 1822)

Burg appears in the Borough tax records from 1815 through 1819. He was a son of the innkeeper John Burg. Although he is said to have made clocks, none are known to survive.

Burrowes, Isaac Bradin. *Strasburg*. (d. 1837)

Burrowes was born in Killigoan, County Cavan, Ireland. He was a younger brother of clockmaker Thomas B. Burrowes. He makes his first appearance in the tax records as early as 1797. From 1797 through 1809 he was listed as "clockmaker." In 1810 he was serving as constable. In 1810 he accompanied his brother back to Ireland. They were in Quebec from 1817 to 1822 and then returned once again to Ireland. In 1825 they came back to Lancaster County. At least one clock is reported to have survived.

Burrowes, Thomas Bradin. *Strasburg*. (b. Oct. 24, 1770, d. 1839)

Burrowes was born in County Cavan, Ireland. He studied for, but never entered, the Episcopal ministry. In 1784 he came to Delaware, where he married in 1786, and is believed to have settled in Lancaster County in 1787. He appears in the town tax records of Strasburg as early as 1797, and his home and shop were located on the southwest corner of the square. In 1810 he returned to Ireland with his family to collect his share of an estate. In 1817 he took them to Quebec, Canada, where they remained until he journeyed once again to Ireland in 1822. In 1825 he brought his family back to Lancaster County and settled at "Oak Hill," now "Christ's Home For Children," in the town of Paradise. A younger brother, Isaac, was also a clockmaker. A son, Thomas H., founded the public school system of Pennsylvania.

Carpenter, Anthony. *New Holland*. (b. ca. 1790, d. Jul. 26, 1868)

Carpenter came to America from Prussia. He appears in the Cocalico tax records from 1817 through 1828. His trade listed in 1828 was that of innkeeper. The census of 1860 lists him as "Gentleman." He is said to have started his jewelry business about 1833 at "Carpenter's Hotel." William Fraser was a journeyman to him. Four sons, Anthony Wayne (1814–1869), Henry Augustus (1815–1846), William (1823–1866), and G. B. Porter (1836–1880), are said to have been apprenticed to Philadelphia clockmakers. Only Anthony Wayne, however, has a known history of clockmaking.

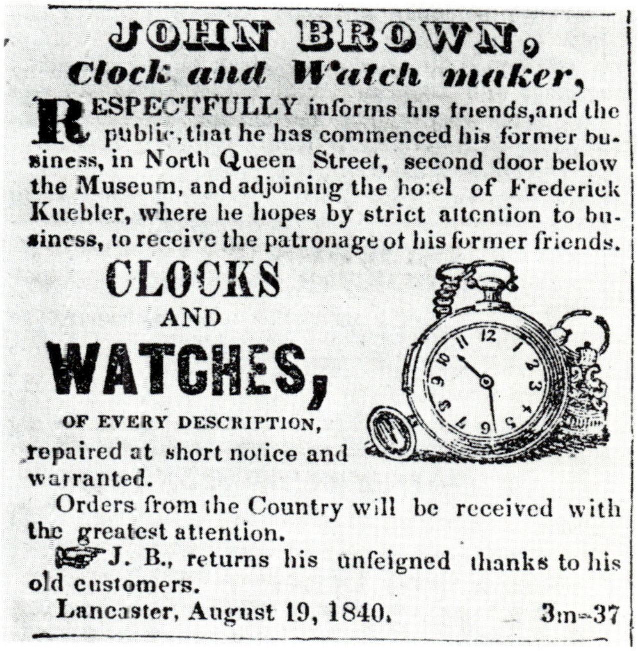

1-2. John Brown ad, *Lancaster Examiner and Herald*, Aug. 19, 1840.

1-3. Thomas Bradin Burrowes by Jacob Eichholtz. (Private collection of a descendant; Photo courtesy of Lancaster Newspapers, Inc.)

Carpenter, Anthony Wayne. *New Holland.* (b. Jan. 1, 1814, d. Jun. 19, 1869)

One of four sons of Anthony and Catharine Carpenter who are said to have been apprenticed to Philadelphia clockmakers, A. W. Carpenter started making clocks about 1834. In 1837 he moved to Burlington in the territory of Wisconsin (now in the state of Iowa) where he established what was said to be the oldest jewelry store in the northwest. He died while on a visit to New Holland and is buried in the Trinity Lutheran Cemetery there.

Cope, John. *Lancaster.* (b. 1762)

Cope was a son of plasterer Caleb Cope (a trustee of the Friends' meeting house and burgess of the Borough in 1774). Major Andre, who had been captured by General Montgomery during the Quebec expedition of 1775, was one of the many prisoners sent to Lancaster. Andre was quartered in Cope's house with other officers and during his stay gave John, then thirteen, drawing lessons. In 1785, John Cope appeared in the tax records as a clockmaker. In 1786 he was listed as a silversmith. From 1787 through 1789 he once again was listed as clockmaker. In 1788, when he purchased a lot in Columbia, he was recorded as watchmaker. The tax list of 1790 for Lancaster Borough states that he was "gone." It is believed that he moved to Richmond, Virginia, for a John Cope advertised in 1792 that he cleaned and repaired musical instruments, plain clocks, and watches. No clocks bearing his name have been located.

Davis, Gabriel. *Manheim.*

Found in the tax records from 1779 through 1781. No clocks have been found bearing his name.

Davis, John. *New Holland.* (b. ca. 1860)

Davis's first appearance in the Earl Township tax records in 1783 lists him as a silversmith. From 1797 through 1805 he was listed as a clockmaker.

Dickey, Thomas. *Marietta.*

Dickey's name is first noted in the tax list of Waterford, Donegal Township, in 1813. By the following year, the name of the town had been changed to Marietta. Dickey remained there at least through 1824. He owned two houses, one on "Market Street" and another on a back street. He later moved to Harrisburg. Dickey's clocks are rare, and none have been found with "Waterford" on the dial.

Eberman, George David. *Lancaster.* (b. Mar. 13, 1818)

A son of clockmaker Joseph Eberman, George started business in 1839. His shop was on the east side of the second square of North Queen Street and his home was in the second square of East Chestnut. In 1844 he kept shop next door to the Cross Keys hotel on West King Street. Later directories list him at 28 North Water Street, Orange and Mulberry, and, in 1869, at 41 North Prince Street. During the Civil War he served as a private in the 15th Pennsylvania Regiment (1861) and the 122nd Pennsylvania Regiment (1862–1863). His clock dials are signed "Geo. Eberman" and are numbered with 17 being the highest reported. Number 4 is

1-4. Anthony Wayne Carpenter from a Daguerreotype. (Photo courtesy of a descendant)

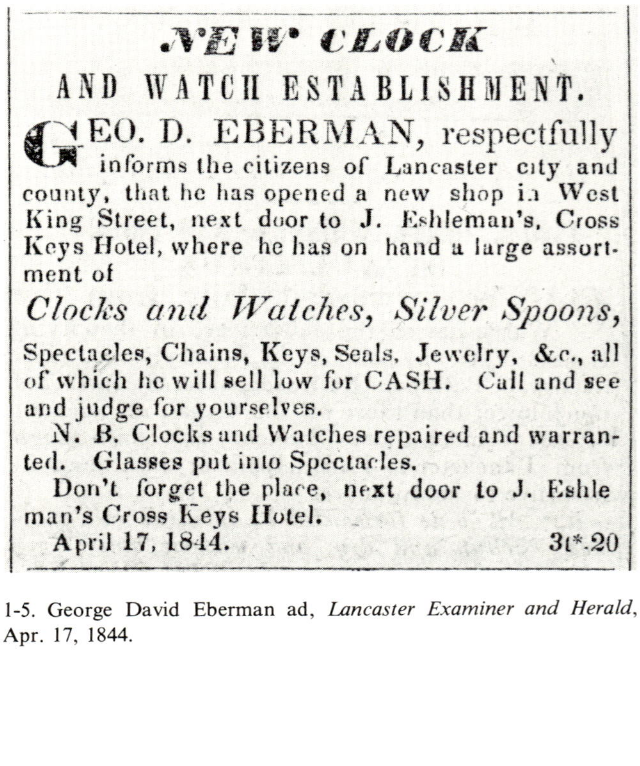

NEW CLOCK
AND WATCH ESTABLISHMENT.

GEO. D. EBERMAN, respectfully informs the citizens of Lancaster city and county, that he has opened a new shop in West King Street, next door to J. Eshleman's, Cross Keys Hotel, where he has on hand a large assortment of

Clocks and Watches, Silver Spoons,

Spectacles, Chains, Keys, Seals, Jewelry, &c., all of which he will sell low for CASH. Call and see and judge for yourselves.

N. B. Clocks and Watches repaired and warranted. Glasses put into Spectacles.

Don't forget the place, next door to J. Eshleman's Cross Keys Hotel.

April 17, 1844. 3t*.20

1-5. George David Eberman ad, *Lancaster Examiner and Herald*, Apr. 17, 1844.

an eight-day movement with a strapped-on eight-bell quarter-hour chime movement.

Eberman, Jacob. *Lancaster.* (b. Feb. 4, 1773)

Jacob was one of two sons of clockmaker John Eberman who followed in the trade. His career was short, however, for the year that he went into business (1796) he lost a hand while helping his father make repairs to the courthouse clock. His clocks are, therefore, rare. In succeeding years he was a grocer, storekeeper, and tobacconist. He served as borough clerk and county treasurer. He moved away from Lancaster about 1815.

Eberman, John, Jr. *Lancaster.* (b. Nov. 2, 1749, d. Jan. 15, 1835)

Eberman was a son of the smith and soap boiler John Eberman (Everman). He worked as a clockmaker from 1772 to 1807. His first shop was on North Queen Street opposite the sign of the "Black Bear." He succeeded George Hoff (Sr.), to whom he had possibly been apprenticed, as keeper of the courthouse clock in 1772, a job which he held for over thirty years. He was commissioned around 1785 to make a second courthouse clock to replace the first one which had been destroyed with the building by fire in 1784. He received £550 for the job. The clock served through 1898 and can now be seen in the museum of the Lancaster County Historical Society. Eberman served as justice of the peace, clerk of the orphans' court, chief burgess of the Borough (1805 and 1806), secretary/treasurer of the Union Fire Company, and was associated with the Lancaster branch of the Bank of Pennsylvania. From 1806 he had the appellation "Esquire." Sons Jacob and Joseph and grandson George David Eberman followed in the trade. Another son, John, was not a clockmaker as previously believed, but served as the Farmers Trust Company's third cashier (treasurer) from 1815 until a few months before his death in 1846. Eberman made thirty-hour and eight-day clocks and at least one musical clock. The inventory of his estate included an eight-day chiming clock valued at $25.00. He used both brass and white dials and signed his name initially "John Eberman, Junior" and finally "John Eberman." One clock carries the number 416 on its dial.

Eberman, John Gottlieb. *Lancaster.* (b. May 28, 1759, d. Mar. 5, 1801)

Gottlieb Eberman was a son of the smith and soap boiler John Eberman, and a brother of clockmaker John Eberman, Jr. He appears in the Borough tax records for the years 1781 through 1801. It is probable that his shop was in his home on King Street. From 1785 to 1787 he owned the "Indian Queen" tavern on East King Street above Church Street. The inventory of his estate includes clockmaking tools (£11 5s.), two clockmaker's signs (5s. 7d.), and a wire drawing machine (3s. 6d.). The dials of his clocks bear the signature "Gottlieb Eberman."

Eberman, John Joseph. *Lancaster.* (b. Dec. 2, 1785, d. Mar. 1, 1860)

Joseph Eberman was a son of clockmaker John Eberman, Jr. He was in business from 1808. The directory of 1843 lists him as "J. Eberman & Son" and places his shop on the east side of the third block of North Queen Street near the railroad station. Son George D. became a clockmaker and son Charles F., a watchmaker. Dials are signed "Joseph Eberman," "Joh Eberman," and "J. Eberman." Early clocks use rack-hook locking possibly indicating that he was apprenticed to someone other than his father. Later clock movements were made elsewhere and have such names as "J. E. Bagnall" (Birmingham, England ca. 1850) and "Wm. Vale—Litchfield" stamped into the front plate and use gathering pallet locking.

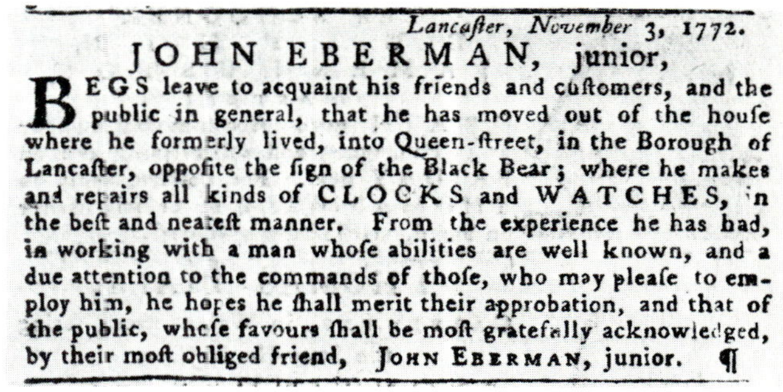

Lancaſter, November 3, 1772.

JOHN EBERMAN, junior,

BEGS leave to acquaint his friends and cuſtomers, and the public in general, that he has moved out of the houſe where he formerly lived, into Queen-ſtreet, in the Borough of Lancaſter, oppoſite the ſign of the Black Bear; where he makes and repairs all kinds of CLOCKS and WATCHES, in the beſt and neateſt manner. From the experience he has had, in working with a man whoſe abilities are well known, and a due attention to the commands of thoſe, who may pleaſe to employ him, he hopes he ſhall merit their approbation, and that of the public, whoſe favours ſhall be moſt gratefully acknowledged, by their moſt obliged friend, JOHN EBERMAN, junior. ¶

1-6. John Eberman, Jr. ad, *Pennsylvania Gazette.* (Philadelphia) Nov. 11, 1772.

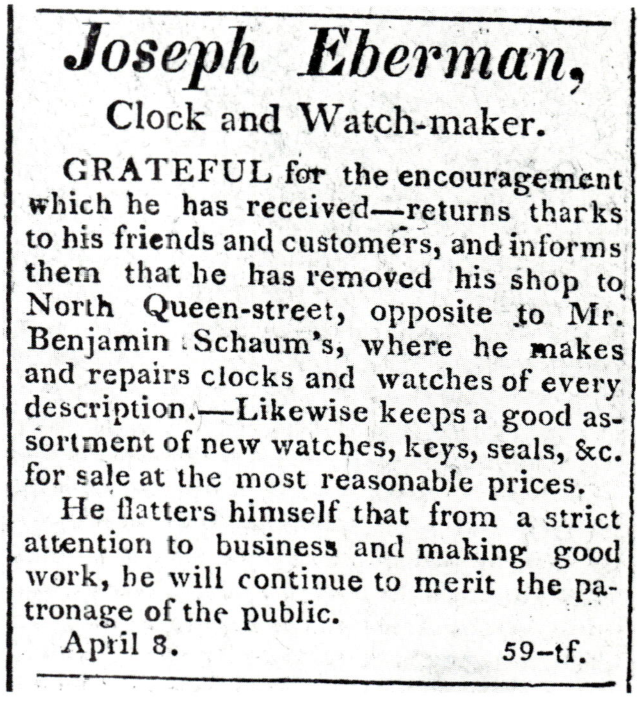

Joseph Eberman,

Clock and Watch-maker.

GRATEFUL for the encouragement which he has received—returns thanks to his friends and customers, and informs them that he has removed his shop to North Queen-street, opposite to Mr. Benjamin Schaum's, where he makes and repairs clocks and watches of every description.—Likewise keeps a good assortment of new watches, keys, seals, &c. for sale at the most reasonable prices.

He flatters himself that from a strict attention to business and making good work, he will continue to merit the patronage of the public.

April 8. 59–tf.

1-7. John Joseph Eberman ad, *Lancaster Journal*, Apr. 8, 1813.

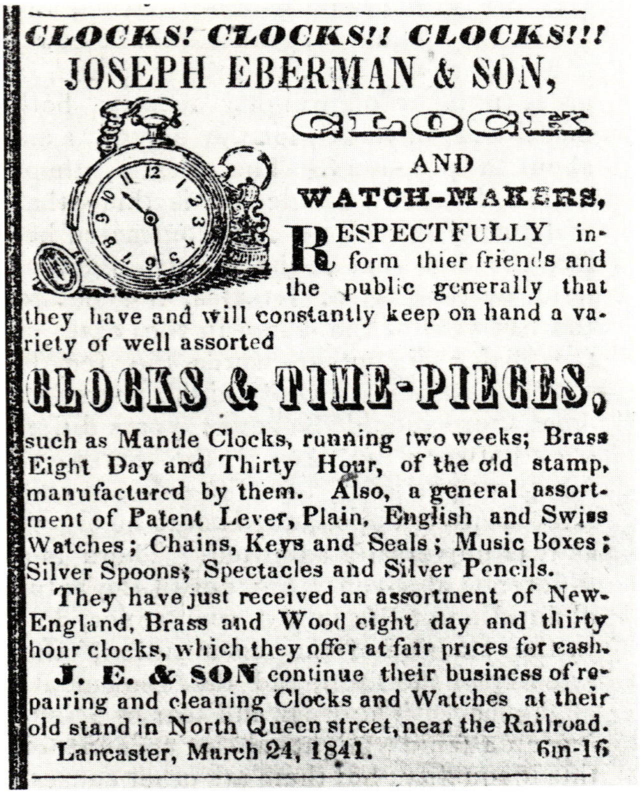

1-8. John Joseph Eberman & Son ad, *Lancaster Examiner and Herald*, Mar. 24, 1841.

Eby, Christian. *Manheim.* (d. 1803)

Eby appears in the town tax records from 1793 through 1803. His shop was located one half a block north of the market square on the west side of Prussian Street. According to legend, he worked with fellow clockmaker Samuel C. Stauffer for a time. Eby's relationship, if any, to clockmaker Jacob Eby, also of Manheim, has not been confirmed although it is believed that they were brothers.

Eby, Jacob. *Manheim.*

Town tax records list Eby from 1803 through 1828. He is believed to have been a brother of Christian Eby, also a clockmaker in Manheim, although this relationship has not been confirmed.

Erb, John. *New Providence* and *Conestoga Center.* (b. 1814, d. ca. 1860)

John Erb was apprenticed to Joseph Bowman, Jr. of Strasburg Borough in 1830 (see his indenture in the beginning of this section). After finishing his term of five years and two months, he went into business in New Providence. He is also noted working in Conestoga Township as early as 1843 and as late as 1860. His final occupation was that of "merchant." His movements are quite similar to those of A. W. Baldwin to whom Joseph Bowman, Jr. had been apprenticed.

Esterle, John. *Maytown* and *New Holland.* (b. ca. 1790)

Name variants are Esterly, Esterlie, and Esterli. John was born in Klingerstown, Pennsylvania, the son of Jacob and Rosina Ried Esterle. During his youth, his family moved to Maytown where he first appears in the tax records of 1813. In 1814 and 1815 he was a deacon of the Reformed Church. In addition to his shop on Main Street, he owned two unimproved lots: one in Bainbridge and one in Centerville. He married Magdalina Merck (Mark) of Lebanon and in 1818 moved to New Holland in Earl Township. In 1828 he was sharing a shop with watchmaker William M'Conaghy. Esterle appears in the Earl Township tax records as late as 1835. According to the family history, he died in Hollidaysburg, Blair County. Some of his Maytown clocks are numbered, one being number 12.

Findler, David B. *Columbia.* (b. 1816)

Findler came to America from Germany and is listed in the census of 1850. No clocks survive.

Fisher, George. *Lancaster.*

Advertised in the *Lancaster Journal* during the summer of 1799 that he had "recommenced the Clock and watchmaking business at his house in King Street, West of the Court House and sign of the King of Prussia." From 1796 to 1799 he had been co-owner of the same King of Prussia tavern with Peter Gonter. Borough tax records continued to list Fisher as an innkeeper. In 1807 and 1808 he appears in the Drumore Township tax lists as a clockmaker and owner of a distillery. He moved to Baltimore, Maryland in 1808 where he kept tavern at the sign of the Black Bear. It is also possible that he is the same George Fisher who later located in Frederick, Maryland. He is believed to have been the eldest son of John Fisher, clockmaker of York Town (now York), Pennsylvania.

1-9. Jacob Eby ad, *Lancaster Journal*, Sept. 20, 1811.

Ford, George. *Lancaster.* (b. Mar. 1, 1773, d. Apr. 17, 1843)

Ford appears as a clockmaker in Borough tax records from 1807 through 1836. His father was Peter Ford, a clockmaker who came to America from Liverpool, England, and settled in York County, Pennsylvania. In 1806 George Ford married Mary Elizabeth Hall, a daughter of Lancaster silversmith Charles Hall. She was also a granddaughter of clockmaker Abrum Laroy. Ford lived on the north side of the first block of West King Street, a few doors from the Hoff shop. He is said to have manufactured nautical, surveyors', and mathematical instruments in addition to clocks. In his will he left his son George Ford, Jr. (1811–1864) all his "Graduating Engines together with such other mathematical instruments and tools" as he might select. George, Jr. had started off as a watchmaker in 1834, but by 1837 he had switched to law. Ford's estate also included two other brick houses on East King Street.

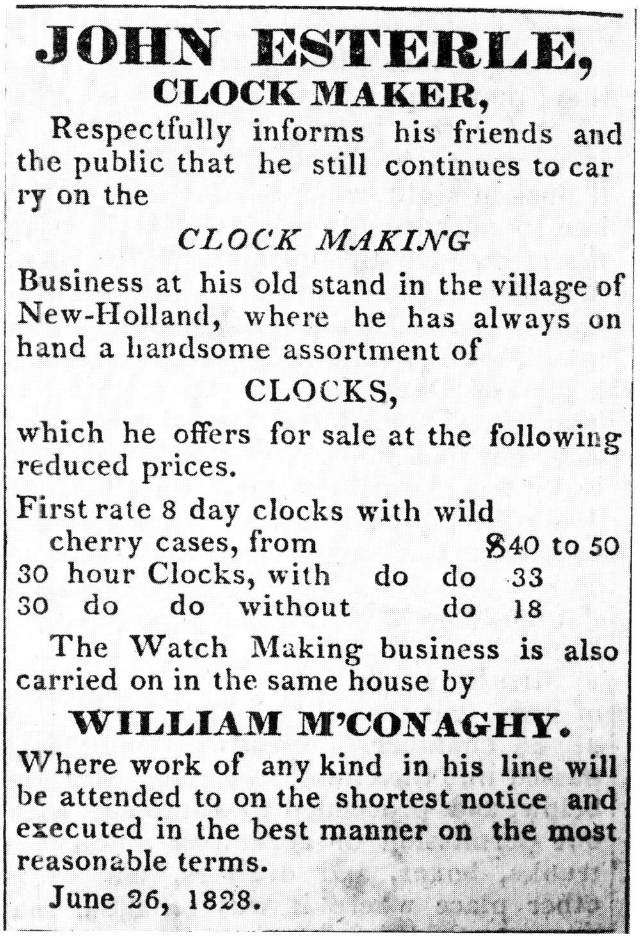

1-10. John Esterle ad, *Anti-Masonic Herald* (New Holland), Jun. 26, 1828.

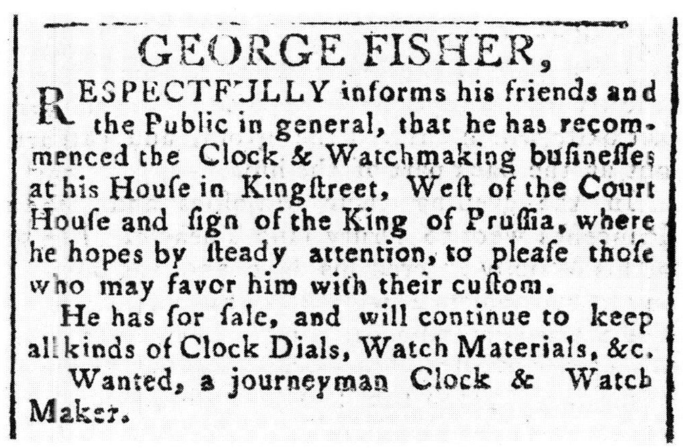

1-11. George Fisher ad, *Lancaster Journal*, Aug. 10, 1799.

Forrer, C. & D. See *Forrer, Christian* and *Forrer, Daniel.*

Forrer, Christian. *Lampeter.* (b. ca. 1737, d. Aug. 28, 1783)

Forrer (Fuhrer) was born in the district of Trachselwald in Switzerland. In 1754, after finishing his apprenticeship to Jean Francois Guillerat of Delemont in the Jura region, he, his brother Daniel, also a clockmaker, and his sister Christina, emigrated to America. They eventually settled in Lampeter. Christian is first listed in the tax records of Lampeter in 1756. He located his shop on the square where he made clocks with his brother which are signed "C. & D. Forrer." He also made a number of clocks, perhaps after Daniel moved away, signed "Christian Forrer." He moved about 1774 to New Cumberland, Newberry Township (now Fairview Township) in York County where he ran a ferry on the Susquehanna River. It is believed that he also kept tavern.

Forrer, Daniel. *Lampeter.*

Forrer (Fuhrer) was born in the Trachselwald district of Switzerland. He and his brother Christian had been apprenticed to Jean Francois Guillerat of Delemont in the Jura. In 1754, Daniel emigrated to America with his brother and sister. They left behind a sizable estate near Mett in the district of Nidau for which they took final settlement in 1757. Daniel appears in the Lampeter tax records of 1758 and 1759. It is believed that he eventually moved to Virginia. While in Lampeter, he made clocks with his brother that carry the signature "C. & D. Forrer." No clocks signed "Daniel Forrer" have been located.

Fraser, William. *New Ephrata.* (b. Feb. 14, 1801, d. Oct. 18, 1877)

Fraser was born in Philadelphia and learned his trade as an apprentice and journeyman to Philadelphia clockmaker Solomon Park from 1814 to 1821. In 1828 he was engaged as a journeyman by Anthony Carpenter in New Holland, Lancaster County. He later was a journeyman to John Esterle, also in New Holland. In 1834 he set up his own business in New Ephrata (now Lincoln), Elizabeth Township, just west of Ephrata. He was succeeded by his son William, Jr. who was in turn succeeded by Samuel K. and William K. Fraser, grandsons of the original William.

Frohm, Sebastian. *Lancaster.*

Appears in the city tax records of 1830 as a clockmaker. The validity of this is doubted, however, as in previous and subsequent years he was always listed as a peddler. Perhaps he peddled the scourge of the Lancaster County clockmaker: the Yankee clock. No known clocks survive.

Gaillard, Peter. *Lancaster.*

Gaillard appears in the Borough tax records from 1811 through 1815. He is believed to be the same P. Gaillard who worked as a clock and watchmaker in Reading, Pennsylvania between 1792 and 1810 and advertised that he had previously practiced his trade in France for many years. No clocks are known to survive.

Getz, Ben. *Lancaster.* (ca. 1785)

A white dial, cherry-cased tall clock signed "Ben Getz/Lancaster" was advertised in the magazine *Antiques* in 1976. The brass movement was described as having lantern pinions, which are indicative of Pennsylvania manufacture. Getz is not listed in the Borough tax records.

Getz, Peter. *Lancaster.* (b. 1764, d. Dec. 29, 1809)

Better known as one of the Borough's finest silversmiths, Getz was one of nine to be listed as clockmaker in the tax list of 1789. In subsequent years his trade was given as silversmith. It is believed that in 1789 the Burgesses of the Borough padded the tax lists with clockmakers in an attempt to demonstrate that Lancaster was a thriving community worthy of becoming the capital of the new nation. In 1796 he advertised fire engines for sale. When he died, his obituary stated that he was an improver of the printing press which was constructed with rollers instead of a screw. No known clocks exist.

Gibble, Lewis. *Manheim.* (b. 1798, d. 1854)

Gibble is listed in the Warwick Township tax records starting in 1822. In 1828 he was listed as a silversmith. He settled in Manheim Borough prior to 1850. No clocks by him are known to survive.

Gorgas, Jacob. *Ephrata.* (b. Aug. 9, 1728, d. Mar. 21, 1798.

Gorgas was born in Germantown, Pennsylvania, a son of John and Psyche Rittenhouse Gorgas. He was a first cousin of Philadelphia's famous clockmaker and astronomer David Rittenhouse (1732–1796). Before becoming a clockmaker, Gorgas had been a skin-dresser. His first appearance in the Cocalico tax records was in 1763, the year he is believed to have married Christina Mack. It is likely that he had arrived in the County earlier with his older brother Benjamin, who is noted in the tax records of 1758. Gorgas owned various pieces of property in the Ephrata area and finally settled in a house that he had built on the north side of Ephrata's Main Street, not far from the Cloister. The house still stands. He is believed to be the Jacob Gorgas who published a catalog of engraved work and dials. He served as sergeant in Capt. John Jones' Company, Col. Peter Grubb's Battalion of the Lancaster Militia that was organized in August 1776. Three sons, Jacob, Solomon, and Joseph, followed in the trade, as did Solomon's son William Rittenhouse Gorgas.

Gorgas, Jacob (Jr.). *Ephrata* and *Hinkletown.* (b. ca. 1763, d. 1828)

The eldest of three sons of clockmaker Jacob Gorgas who followed in their father's trade. He appears in the Cocalico Township tax records from 1786 through 1791 and then in the Earl Township records from 1791 through 1822. A letter from him dated 1826, ordering clock parts, indicates that he had moved to Elizabethtown (Mount Joy Township). He was still there in 1828. Since he did not follow his name with "Junior," it is quite difficult to tell which clocks made in Ephrata were made by the father, and which by the son, unless one studies the movements carefully. Both are believed to have used brass and white dials.

Gorgas, Joseph. *Ephrata.* (b. 1770)

The fourth son of Jacob Gorgas (Sr.), Joseph was born in Ephrata and worked there from 1792 through 1805. In the Cocalico Township tax records of 1800 he is listed as having a saw mill. He moved about 1806 to Mount Joy Township, near Elizabethtown, where he ran the Running Pump Inn until about 1815. He is said to have made clocks there. He eventually moved to Harrisburg. It is possible that Henry Ober was apprenticed to him. Like his brother Solomon, he did not

1-12. Jacob Gorgas (Sr.) House, Main Street, Ephrata.

normally have his town's name painted on his clock dials. His clocks are extremely rare, and until recently it was thought that none had survived. Gorgas married Sophia Atkinson who is believed to have been a daughter of Lancaster Borough clockmakers Wilton and Anna Maria Laroy Atkinson. A son, Josiah, became a brigadier general in the Confederate Army and eventually served as president of the University of Alabama. Josiah's son William Crawford Gorgas successfully combated yellow fever in the Canal Zone and became surgeon general of the United States Army. Gorgas's movements have been noted with both gathering pallet and rack hook locking. An early movement has lantern pinions although he normally used drawn pinions.

Gorgas, Solomon. *Ephrata.* (b. Jan. 22, 1764, d. Sept. 21, 1838)

The second son of Jacob Gorgas (Sr.), Solomon was born in Ephrata and worked there as a clockmaker from 1787 through 1804. He was also listed as "a retailer of goods." He moved to Lower Allen Township, Cumberland County where he kept the first tavern and store in that area. He served two terms as a member of the state legislature (1821 and 1822). A son, William Rittenhouse Gorgas (1806–1892), followed in the clockmaking trade for a time.

Grosch, Samuel S. *Marietta* and *Mount Joy.* (b. ca. 1800, d. ca. 1856)

Grosch first appears in the 1817 tax records of Donegal Township. He is listed as being in partnership with a Mathias Tschudy about whom no further information is available (a Mathias Tschudy of Lititz was a contemporary whose trade was woolen weaver and who specialized in chip hats). The partnership continued through 1824, with their holdings fluctuating from one house and one lot, to one house and eleven lots. In 1828 the township list of taxables records Grosch as a clockmaker. The census of 1830 shows that he was still in Marietta and records his household as consisting of himself, one female under five, and another female, probably his wife, between the ages of twenty and thirty. The next time he was recorded was in the census of 1850 where he is listed as a fifty year old Lancaster County born watchmaker living in Mount Joy. The only recorded deed concerns settlement of his estate in 1856. A recently discovered fragmented watch-paper bears the following printed message: "Samuel S. (Gr)osch, Clock and Watchmaker, Walnut (Street), Marietta." A line has been inked through the address and "Mount Joy" penned in. One of the few existing Lancaster County shelf clocks bears Grosch's name (Figs. 2-55 and 4-86). An eight-day tall clock has also been reported.

Gunkle, John. See *Kunkel, John.*

Hall, Christian. *Lititz.* (b. May 10, 1775, d. June 30, 1848)

Hall was born in the Moravian community of Bethlehem, Pennsylvania. He was a step-son of the famed organ builder David Tannenberg. Hall first appears in Lititz (Warwick Township) tax records of 1799 as an innkeeper. From 1803 to 1822 he ran the "Anchor Hotel" (later renamed the "Lititz Springs Hotel" and now known as the "General Sutter Inn"). He also served as town postmaster from 1806 to 1822. The first mention of him as a clockmaker in the tax records was in 1828. He eventually ran a small grocery store. He owned property in the village of Warwick as well as in the "closed" Moravian town of Lititz. He left his clockmaking tools to a son William Henry Hall who made at least one clock. Hall's dials carry his name with the town's name spelled "Litiz" or, occasionally, the new spelling "Lititz."

Hall, William Henry. *Lititz.* (b. Sept. 9, 1809, d. May 19, 1868)

Hall was a son of clockmaker Christian Hall and his wife Catherine Protzman. He was a painter by trade and a tower clock dial that he painted for the Ephrata Academy still survives. He also served as teacher at both the Nazareth Hall School and the Lititz Academy. In addition, he was organist for the Moravian congregation.

Hanebarger, Peter.

Appears in the Rapho Township tax record of 1807. No clocks by him have been located.

Hay, Samuel. *Lancaster.*

Hay is noted in the Borough tax records of 1789 only. This was the year that the borough fathers were trying to get the Federal Government to select Lancaster as the nation's capital. No clocks bearing Hay's name are known to exist.

Heinselman, Hieronymus. *Manheim.* (b. Aug. 9, 1730, d. Nov. 25, 1796)

Heinselman, a Swiss immigrant, appears in the Manheim Town tax records as early as 1770. Although his occupation was listed as tavern keeper, it is believed that he was also a clockmaker. His clocks are said to be signed "H. H." and "Heintzelman."

Heinselman, John Conrad. *Manheim.* (b. Aug. 27, 1766, d. Sept. 1, 1804)

A son of Hieronymus Heinselman (Heintzelman), he appears in the Rapho Township tax records from 1792 through 1803. Like many clockmakers, he was also an innkeeper and is so listed in the tax records. He is credited with building the first inn in Manheim Town, the Black Horse.

Heisely, Frederick. *Lancaster.* (b. 1759, d. Dec. 3, 1839)

Heisely was born in Lancaster and is said to have been apprenticed to George Hoff (Sr.) whose eldest daughter, Catharine Julianna, he married in 1783. He was listed as a watchmaker in the Borough in 1783. From approximately 1784 to 1795, he was engaged in the manufacture of surveyors' instruments in Fredericktown, Maryland. In January 1796, he entered into a partnership with George Hoff at Hoff's shop in the first block of West King Street. Within ten months the partnership had dissolved and Heisely removed to the third block of West King Street where he carried on clockmaking as well as the surveyors' instrument business. He moved about 1798 to Harrisburg where he and his sons George J. and Frederick A. conducted business. They made a tower clock for the Columbia Borough town hall (ca. 1828). Another tower clock by them is to be found in the Smithsonian Institution.

Hoff, Frederick George. *Lancaster.* (b. 1810, d. Mar. 18, 1852)

Hoff was a son of clockmaker John Hoff and a grandson of George Hoff (Sr.). He is first noted in the tax records of 1831. In 1834 he advertised that he had commenced the clock and watchmaking business in West King Street. This was probably the home and shop of his late father. He acted as agent for the Pittsburgh gunsmiths E. & F. Faber in 1835. Sometime after 1837 he moved to Aurora, Indiana where he died in 1852. No known clocks survive.

Hoff, John. *Lancaster.* (b. Sept. 2, 1776, d. Sept. 15, 1818)

The eldest of George Hoff's three clockmaking sons, Hoff started his own business in 1799. In the following sixteen years he made 103 clocks. In addition to the standard thirty-hour and eight-day tall clocks, he also made four quarter-hour chiming clocks and one ten-tune musical clock, as well as spring powered shelf and wall clocks. Orders were filled for the surrounding counties and as far away as Niagara and North Carolina. His first shop was on the north side of East King Street next door to the Pennsylvania Arms tavern. In 1807 he bought his father's home and shop on the north side of the first block of West King Street (now a parking lot). He served as an elder of the Evangelical Church of the Holy Trinity, temporary cashier (treasurer) of the Farmers Trust Company, assistant borough burgess, and finally, in 1816, burgess. He died of typhus in 1818.

Hoff, John George. *Lancaster.* (b. Oct. 22, 1733, d. Aug. 18, 1816)

Hoff learned the art of clockmaking from his father-in-law, George Schnertzel, in Grunstadt, the Palatinate, Germany. According to tradition, he and his family arrived in Philadelphia, Pennsylvania on August 24, 1765. His name does not surface in Lancaster, however, until 1769. It is possible that he became indentured in order to pay for his family's passage. It is also possible that Hoff was invited to settle in the Borough following the death of the Borough's sole clockmaker, Rudy Stoner, in April of the same year. Hoff purchased a house and half a lot on the north side of the first block of West King

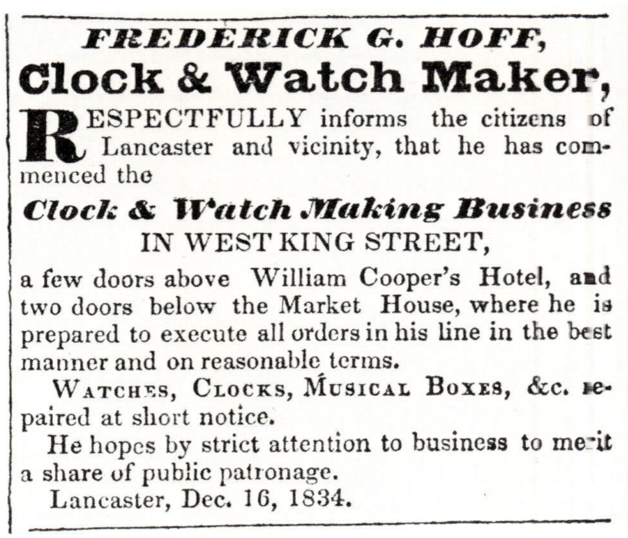

FREDERICK G. HOFF,

Clock & Watch Maker,

RESPECTFULLY informs the citizens of Lancaster and vicinity, that he has commenced the

Clock & Watch Making Business

IN WEST KING STREET,

a few doors above William Cooper's Hotel, and two doors below the Market House, where he is prepared to execute all orders in his line in the best manner and on reasonable terms.

WATCHES, CLOCKS, MUSICAL BOXES, &c. repaired at short notice.

He hopes by strict attention to business to merit a share of public patronage.

Lancaster, Dec. 16, 1834.

1-13. Frederick George Hoff ad, *Lancaster Examiner and Herald* Dec. 16, 1834.

1-14. John Hoff by Jacob Eichholtz. (Private collection of Mrs. Anthony J. Wierschem, a descendant; photo courtesy of Frick Art Reference Library)

Street, just down from the courthouse. The site is now a parking lot. He immediately assumed charge of the courthouse clock. It appears likely that John Eberman finished out his apprenticeship with Hoff. In any case, Eberman relieved Hoff of the courthouse clock duties when he finished his term and went into business in 1772. In January 1796, Hoff entered into a partnership with his son-in-law, Frederick Heisely. Heisely brought with him his skills as a maker of mathematical and surveyors' instruments. The partnership lasted only until November of the same year, and Heisely removed to the third block of West King Street where he went into competition with Hoff. Hoff served as a private in the Jasper Yeates Company of Colonel Matthias Slough's Battalion of Associates in 1776. He was an assistant to the Borough burgesses in 1778, 1779, and 1785. He was also a member of the Friendship Fire Company. In 1807 he made his will and sold his house to his clockmaker son John. Two other sons, John Jacob and John George, Jr., were also clockmakers. A grandson, Frederick G. Hoff, became a watchmaker. Having come to America as a full-fledged clockmaker, Hoff continued to make his clock movements in the German tradition, using lantern pinions and open-ended pinions, iron movement pillars, cock-and-loop pendulum suspension, and rack hook strike train locking. His early dials were painted iron with attached pewter chapter ring and spandrels. It should be noted that all of these dials having their original parts have identical spandrel castings. The dials are screwed to the extended movement pillars. Eight-day clocks normally have asymmetrical winding arbor holes in the dials. Hoff made modifications to his movement design over the years, working in some of the better English features. His dials carry the signature "George Hoff," It is possible that many later white (painted) dial clocks with the same signature were in fact made by his son John George Hoff, Jr.

JOHN HOFF,
Watch and Clock Maker.

RESPECTFULLY informs his friends and the Public in general, that he has commenced bufinefs in the Houfe formerly occupied by Mr. John Gallagher, in eaft King-ftreet, Lancafter, next door to Mr. Slaymaker's tavern, where he will carry on the Clock and Watch making bufineffes in all their branches. Thofe who favor him with their cuftom may rely upon the ftricteft attention, punctuality and difpatch.

N. B. Clocks and Watches warranted.
July 5.

1-15 John Hoff ad, *Lancaster Journal*, Jul. 5, 1800.

JOHN HOFF,
CLOCK & WATCH-MAKER,

RESPECTFULLY informs his friends and the public that he has removed from his late stand, near the Stage, office of Mr. Slaymaker, to the old stand of his father, next to Mr. Hamilton's Printing-office, between the Court-house and Jail, and opposite Daniel Witmer's tavern; where he intends carrying on his business with care and attention, as formerly; and likewise will make, and always have on hand, Surveyor's Compasses, and other Mathematical Instruments, of every description. He keeps a constant supply of all kinds of new Watches, Clock-dials, and all other clock and watch materials. Any work done will be warranted.

He is thankful for the favors received from a generous public, and hopes to merit a continuance.

He keeps Clock-glass, which he cuts to all sizes, to fit clock-cases.

Orders from the country will be punctually attended to.

Lancaster, March 27, 1807. 43—

1-15 John Hoff ad, *Lancaster Journal*, Mar. 27, 1807.

1-17. John George Hoff by Jacob Eichholtz. (Private collection of Mrs. Anthony J. Wierschem, a descendant; Photo courtesy of Lancaster County Historical Society)

Hoff, John George, Jr. *Lancaster*. (b. Jun. 9, 1788, d. Jan. 25, 1822)

One of three clockmaking sons of George Hoff, he first appears in the Borough tax records of 1811. The location of his first shop is unknown, but in 1818 he relocated to the north side of the first block of East King Street, two doors from the stage office. In addition to clocks and watches, he also carried surveyors' compasses and mathematical instruments. Following his death in 1822 at the age of thirty-three, his widow brought in clock and watchmakers from Philadelphia in an attempt to keep the business going. It is not known how successful she was, however, tax records carry her name through 1837. Hoff's clock dials are signed "George Hoff, Jur./Lancaster." It is probable that following his father's death in 1816 (and perhaps as early as 1807) he dropped the "Jur." from his signature.

1-18. John George Hoff, Jr. by Jacob Eichholtz. (Photo courtesy of a descendant)

George Hoff, Jun.

CLOCK AND WATCH-MAKER.

RESPECTFULLY informs his friends, and the public in general, that he has commenced the CLOCK & WATCH-MAKING business in East King-street, Lancaster, next door to Mr. Slaymaker's Stage-office, and nearly opposite to Mr. Eichholtz's Tavern, sign of the Bull, where he intends carrying on his trade with care and attention, and hopes to merit the fovours of a generous public—He will always have on hand, Surveyor's Compasses, and other Mathematical Instruments, of every description; and likewise keeps a constant supply of all kinds of new Watches, Chains, Seals, Keys, and all other Clock and Watch-Meterials. Any work done will be warranted.

☞ Orders from the country will be punctually attended to.

November 13, 1812, 35—tf.

1-19. John George Hoff, Jr. ad, *Lancaster Journal*, Nov. 13, 1812.

Hoff, John Jacob. *Lancaster*. (b. Feb. 4, 1784)

Hoff first appeared in the Borough tax records of 1807 when he was listed as a clockmaker. This was the same year that he moved from the home of his famous clockmaking father George Hoff (Sr.) on West King Street, to a shop next door to the stage office on East King. Although he advertised himself as a clock and watchmaker, he apparently only sold and serviced watches. He also carried chains, seals, and keys. He left Lancaster in 1810 and may have moved to Reading, Pennsylvania. If he made any clocks, none are known to survive.

Houston, James. *Lancaster*. (b. ca. 1780)

Houston was listed as a clockmaker in the tax records of 1808 and 1809. In 1810 he apparently gave up the trade to devote full time to his position as treasurer

Notice.

ALL persons indebted to the estate of

GEORGE HOFF,

Deceased, are requested to make payment, and all who have claims to present them to

MARGARET HOFF, *Executrix*.

MRS. HOFF,

HAVING procured good workmen from Philadelphia, continues the Clock and Watch making business, and solicits a share of public patronage. She has on hand a good assortment of Clocks and Watches, which are offered for sale on reasonable terms.—Also, a large Clock suitable for some public building.

Mrs. Hoff has also and intends constantly to keep a general assortment of Leghorn & Straw

Bonnets,

Together with every other article in the Millinery line.

Feb. 19 29—tf

1-20. Margaret Hager Hoff (Mrs. John George Hoff, Jr.) ad, *Lancaster Journal*, Feb. 19, 1822.

of the Lancaster Borough branch of the Bank of Pennsylvania, a position he had held from 1803. No clocks by Houston are known to have survived.

Huber, Christian. *Reamstown.* (d. 1789)

Huber is listed in the Cocalico tax records of 1785 through 1787. The inventory of his estate shortly after his death listed his tools at £10 10s., two clocks at £15 14s. 9d., a clock with case at £4 15s., and a watch at £3 10s.

Hunchberger, Isaac. (b. 1804)

Appears in the West Earl tax records of 1849 and 1850. No other information is available.

Hunsecker, Jacob. *Marietta.* (b. 1809)

Appears in the East Donegal Township tax records from 1842 through 1850. No known clocks by Hunsecker survive.

Jameson, Jacob. *Columbia.* (d. ca. 1830)

Jameson is believed to have had a clockmaking shop as early as 1818 in Columbia on Front Street between Locust and Walnut Streets although his first appearance in the West Hempfield tax records was in 1820. He is last seen in the tax record of 1822. By 1828 he had left and is next noted in Dayton, Ohio and finally in Springfield, Ohio. In 1880 it was reported that a clock made by him, which would run a month, was located in Harrisburg.

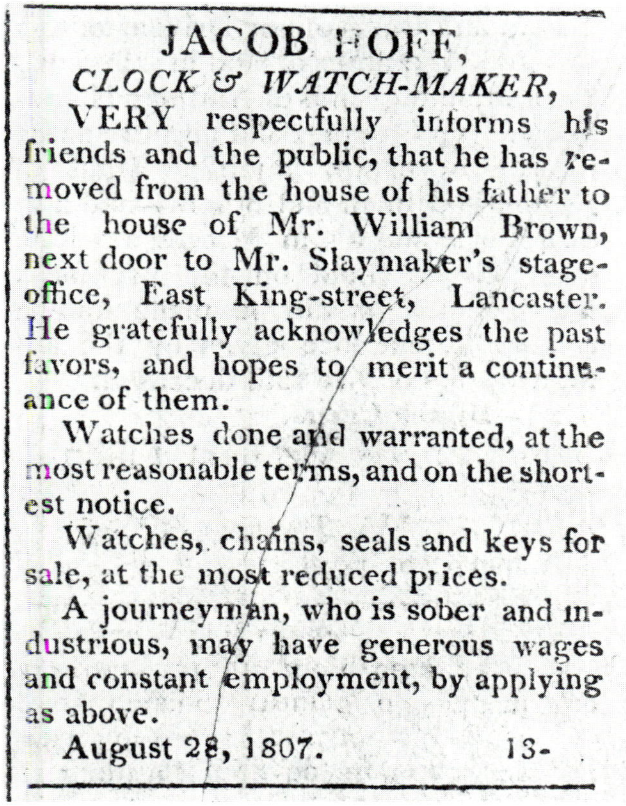
JACOB HOFF,

CLOCK & WATCH-MAKER,

VERY respectfully informs his friends and the public, that he has removed from the house of his father to the house of Mr. William Brown, next door to Mr. Slaymaker's stage-office, East King-street, Lancaster. He gratefully acknowledges the past favors, and hopes to merit a continuance of them.

Watches done and warranted, at the most reasonable terms, and on the shortest notice.

Watches, chains, seals and keys for sale, at the most reduced prices.

A journeyman, who is sober and industrious, may have generous wages and constant employment, by applying as above.

August 28, 1807. 13-

1-21. John Jacob Hoff ad, *Lancaster Journal*, Aug. 28, 1807.

Kammerer, Jacob.

Warwick Township tax records from 1799 to 1818 list a Jacob Kammerer who was a joiner. A thirty-hour tall clock bearing his name but without a town name exists. The movement uses rack hook locking. It is possible that he made the case and had his name painted on the dial.

King, Abraham. *Lancaster.*

Recorded in the Borough tax records of 1756–1759. King is believed to be *Abrum Laroy.*

Kinkead, Alexander. *Lancaster.*

Kinkead is listed in the Borough tax records of 1783 through 1785. He also kept tavern. This may be the same Alexander Kinkead who was a son of clockmaker Joseph Kinkead of Christiana Bridge, Delaware. No Lancaster clocks by this maker have been found.

Kunkel, John. *Trout Run.*

Cocalico tax records show that the German born Kunkel was engaged as a clockmaker from 1813 through 1836. Trout Run became part of the new Ephrata Township when Cocalico was divided into three separate townships in 1838. Kunkel is said to have had a wheel cutting engine. His name also appears as "Gunkle." None of his clocks have been located.

Landry Smith. See *Smith, Alexander Landry.*

Laroy, Abrum. *Lancaster.* (b. May 24, 1701, d. Sept. 24, 1763)

Laroy came to America about 1742 from Soncebozen in the bishopric of Basel, Switzerland. He first appears in Lancaster County records in 1757 when he was paid £1 15s. for three months' care of the courthouse clock. The clock remained in his care until his death in 1763. Son John Peter then took charge of it until his unexpected death in 1765. Laroy's home and shop was located on the west side of the first block of North Queen Street, a few doors from the courthouse. His daughter Salome and her husband, Lancaster Borough silversmith Charles Hall, owned the property until 1795 when they transferred ownership to their son, silversmith David Hall. Laroy's daughter Anna Maria married clockmaker/cutler Wilton Atkinson. The voluminous inventory of Laroy's estate, valued at £929, indicates that he was also a dry-goods merchant.[1] Of horological interest are three chamber clocks and three tall clocks, one of the latter identified as a "Joim Clock"

1. The complete inventory appears in Vivian S. Gerstell's *Silversmiths of Lancaster, Pennsylvania 1730–1850* (Lancaster County Historical Society, 1972)

(possibly a musical clock) valued at £24, two cutting engines, four lathes, and a fusee tool. Laroy signed his name "Abrum Laroy" although his name is often seen spelled *Abraham Le Roy*. No clocks bearing his name are known to have survived. See also *King, Abraham*.

Leinbach, Elias.
Leinbach is first noted in the Earl Township tax lists of 1801 through 1816. He next appears in the Brecknock Township lists from 1845 through 1854. His occupation was given as "rakemaker" in 1852. He possibly lived in Bowmansville, for it was there in 1850 that he took part in the riots against compulsory, or free, schooling. Leinbach was said to be more famous for his belief in witches, hobgoblins, and secret buried treasures than he was for his clocks and brushes.

Leinbach, John. *Reamstown.*
Appears in the Cocalico Township tax records of 1792–1798. He apparently specialized in thirty-hour tall clocks.

Leroy, Abraham. See *Laroy, Abrum*.

Londresmith, Alexander. See *Smith, Alexander Landry*.

Miller, George. *Lancaster*. (d. 1806)
Tax records show that Miller worked as a clock and watchmaker in the Borough from 1800 to 1806. In 1800 he bought lots number 440 and 441 on the east side of Charlotte Street, the former situated at the corner of Lemon Street. No clocks bearing his name are known to survive.

Montandon, Hannah. *Lancaster.*
Born Joanna Shweiker, she became the wife of Borough clockmaker Henry Lewis Montandon. Following her husband's death in 1802, she attempted to keep the business going by forming a partnership with Oliver Roberts. She appears in the Borough tax records only in 1803. By January 1804 she had married Henry Barchet, a baker, and with the settlement of Montandon's estate, she and her new husband apparently moved to Clarksburg, Virginia (now part of West Virginia) where Montandon owned property. In 1824, a Charles M. Barchette (sic) advertised himself as a clock and watchmaker, silversmith, and jeweler, located in a shop recently occupied by his stepbrother Albert Montandon at Mack's Hotel in Clarksburg. The relationship, if any, with New York watchmaker Julien Montandon, is unknown. No clocks made by Hannah and her partner Roberts are known to exist.

Montandon, Henry Lewis. *Lancaster*. (b. ca. 1777, d. Jul. 21, 1802)
Montandon is listed in the Borough tax records from 1797 through 1800. His shop was located on North Queen Street near the courthouse. The inventory of his estate taken after his unexpected death in 1802 indicates that he may have come to Lancaster from what is now West Virginia, for the final item is "300 acres of Land in Harrison County (Virginia)." The inventory, with a total valuation of £758 19s. 72d., also includes completed eight-day clocks, movement castings for eight-day clocks, clock dials, "two Compleat Engines for Cutting Watch Wheels," and various other clock and watch parts. His clock and watchmaker tools were valued at £60. New watches on hand were "Robinson," "Morgan," "Strange," "Hunt," "Stein," "Cumberland," "Jackson," "Cross," and "Gibson," the names being those of eighteeth-century English watch manufacturers. His widow, Hannah, attempted to continue the business in partnership with Oliver Roberts.

Ober, Henry (Jr.). *Elizabethtown.*
Ober may have learned clockmaking from Joseph Gorgas. Ober is first noted in the Mount Joy Township tax record of 1828. Deeds show that he owned property in West Donegal Township by the Conawago Bridge. Letters exist signed by Ober requesting brass, weights, pinions, dials, etc. from the Harrisburg merchants Ogelsby & Pool. He is last noted in the tax record of 1843. In addition to tall clocks, a rare early shelf clock by Ober survives.

Quest, Samuel. *Maytown.* (b. ca. 1795)
According to tax records, Quest first appeared in Maytown in 1817, the year after clockmaker John Esterle had moved on to New Holland. Quest is not noted again until after 1820.

Richwine, Isaac. (b. 1808)
Richwine was a clockmaker located in Paradise Township according to the 1850 census. No known clocks by him survive.

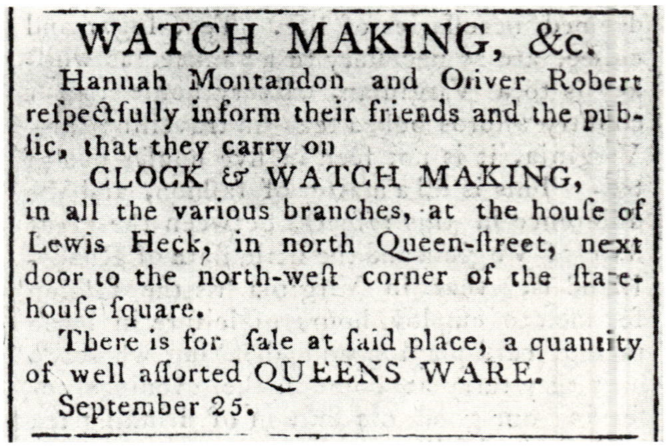
WATCH MAKING, &c.
Hannah Montandon and Oliver Robert reſpectfully inform their friends and the public, that they carry on
CLOCK & WATCH MAKING,
in all the various branches, at the houſe of Lewis Heck, in north Queen-ſtreet, next door to the north-weſt corner of the ſtate-houſe ſquare.
There is for ſale at ſaid place, a quantity of well aſſorted QUEENS WARE.
September 25.

1-22. Joanna (Hannah) Shweiker Montandon (Mrs. Henry Lewis Montandon) ad, *Lancaster Journal*, Sept. 25, 1802.

Roberts, Oliver. *Lancaster.*

Roberts is said to have been a journeyman to Borough clockmaker Henry Lewis Montandon. Following Montandon's death in 1802, his widow, Hannah, attempted to keep the business going with Roberts as a partner. She was successful for eighteen months or so. Roberts was a witness at the settlement of the Montandon estate in January 1804 and is not noted again, nor did he ever appear in the Borough tax records. He is believed to be the "O. Roberts" who was in business in Eaton, Ohio about 1825.

Shreiner, Martin. *Lancaster.* (b. Jan. 23, 1769, d. Feb. 14, 1866)

Martin, a son of Philip and Eva Catherine Boltz Shreiner (Schreiner/Schreyner), was born in a home built by his father on North Queen Street, just north of the Borough's famous Grape Hotel. He is believed to have learned clockmaking from John Eberman. In July 1791 he placed an advertisement in the *Lancaster Zeitung*, a German language newspaper, stating that he had opened a clockmaker's shop. He offered to make house (tall) clocks in either of two styles: the German style with wire pinions, or the English style with drawn pinions. The former sold for 15s. 9d. less than the latter but still carried a ten-year guarantee. Shreiner's clocks carry perhaps the most beautiful of all the signatures ever put on a clock dial. In addition, the dials are numbered, the highest reported being number 431. One of his clocks, number 227, was once owned by General Robert E. Lee. In 1830, Shreiner constructed a fire engine for the city's Sun Fire Company. That same year he turned his clock business over to his sons Martin, Jr. and Philip. He rebuilt the Lititz German-made fire engine, and in 1836 he made an engine for Lancaster's American Fire Engine and Hose Company. He served as engineer for the latter in 1834 and as the company's president in 1841 and 1845. In 1832 an advertisement in the *Lancaster Journal* invited the public to view an orrery which Shreiner had constructed. It showed the motions of the sun, Mercury, Venus, earth, and the moon "with great precision." It also could reproduce the eclipses. Shreiner is buried in the cemetery that he founded in 1836 at the corner of West Chestnut and Mulberry Streets, the same cemetery in which the emancipator Thaddeus Stevens is buried. Shreiner's clock movements are distinguishable by his removal of a trapezoidal piece from the bottom edge of both plates, gathering pallet locking, and, usually, the use of sweep calendar and seconds hands.

Henry B. Montandon

Sack und Haus-Uhr-Macher,

in der Nord Queen Strase in Lancaster.

Hat so eben erhalten und zu verkaufen um die billigsten Preise ein Aßortement englischer Sack-uhren, Maschinen und alle Geräthschaften die zum Uhrmacher gehörig, sind sowohl als
Uhrgesichter von 12 bis 15 Zoll
Sackuhr Zifferblätter.
Goldne, vergoldete und stählerne Zeiger
Sackuhren Gläser von allen Grösen.
Grose und Haarfedern.
Inwendig Ketten.
Auswendige ditto, Schlüssel und Petschafte.
Silberne und vergoldete Pendants u. Bügel, und alle zu Sackuhren gehörige Materialien zu weitläufig zu melden.

Besagter Montandon wünschte einen Gesellen der mit dem Hausuhrmacher Geschäfte schon bekant ist und der Lust hätte das Sackuhr Machen bey ihm in volkommenen zu lernen. Derselbe würde gute Ermunterung bey ihm finden.

Den 27 Dec.

1-23. Henry Lewis Montandon ad, *Lancaster Correspondent*, Dec. 27, 1800.

Martin Schreiner,

Uhrmacher.

Neben des Herrn Adam Reigarts Stohr in der Stadt Lancaster, bittet um Erlaubniß seinen Freunden und dem Publicum für die Ermunterung die er seit dem Anfang seines Geschäfts gefunden, seinen Dank abzustatten; welches ihn in Stand gesetzt hat, dasselbe nach ausgebreiteten Plan und auf solche Bedingungen anzutreiben, welche allgemeine Genehmigung zu sichern wird.

Besagter Schreiner fähret fort, alle Sorten von Hausuhren auf die neueste mode zu verfertigen, er giebt sie um den billigsten Preiß; er machet auch bekannt das er zweierlei Sorten Hausuhren machet, wie man sie gern hat, auf die Deutsche oder auf die Englische Art, die Deutsche Art so genante drathene Trieb, die werden um 15s. und neun pens wohlfeiler verfertiget, als wie die andern, und garantirt solche Hausuhren für zehn Jahr.

Es befinden sich aber Verächter solcher Uhren, aber glaubts ihnen nicht, dann ein jeder Mensch hat sein Verächter, ein gutes Werk machet ein guter Meister, er hat sich auch fürgenommen, daß er vorzüglich die Taschenuhren Reparirt, und stehet das sie ein Jahr gehen, wan sie nicht durch Zufall verbrochen werden.

1-24. Martin Shreiner ad, *Lancaster Zeitung*, Jul. 27, 1791.

Martin Schreiner,
Uhrmacher,
Bittet um Erlaubniß seinen Freunden und dem Publikum bekannt zu machen, daß er seine Werkstatt von der Nordseite nach der Süd-seite der Queens-straße in der Stadt Lancaster, und in dem bequemlichen Hauße wo Henrich Zehmer, und die nächste Thür allwo Herr Docter du Fresne gewohnt verlegt hat; Allwo er alle Sorten von Uhren verfertiget, Sackuhren reparirt und an Bezahlung nimmt er Korn, Holz und Weizen.

1-25. Martin Shreiner ad, *Lancaster Zeitung*, Jun. 6, 1792.

INVITATION.

Every Lady and Gentleman in the city of Lancaster and its vicinity, who has a taste for science and art, is most respectfully invited to call at Mr. King's Hotel, in South Qeen St. andview the most perfect and beautiful Astronomical Instrument ever seen on this side of the Atlantic.

This instrument was constructed by Mr. Martin Schreiner, of this city, under the direction of the subscriber. It is technically called a *Tellurian* and *Lunarian*, and shows the motions of the Suu, Mercury, Venus Earth and Moon, with great percision. By turning a small crank all the Eclipses that ever occurred since our Earth and Moon commenced their celestial journies can be determined or for any future given period.

All persons are freely permitted to examine the instrument, from 9 A. M. to 9 P. M. every day this week, without money or price.

R. DENNIS.

August 6, 1832.

1-26. Martin Shreiner "Tellurian" ad, *Lancaster Journal*, Aug. 10, 1832

Shreiner, Martin (Jr.). *Lancaster*. (b. ca. 1806, d. Mar. 27, 1879)

Martin, Jr. was a son of Lancaster's prolific clockmaker, Martin Shreiner. He started business in 1829 and the following year, with his brother Philip, took over his father's clock business. The partnership lasted until 1838 when Philip moved away. City directories list him at 26 North Queen Street in 1843 and across the street the following years. A son, Henry M., also took up the trade about 1845. While working with his brother Philip, their clocks carried the signature "M. & P. Shreiner" and were numbered. At least forty were made. Just how many clocks, if any, were made following the break-up of the partnership is unknown.

Shreiner, Philip. *Lancaster*, and *Columbia*. (b. Dec. 1808, d. May 25, 1877)

Philip was the younger of two clockmaking sons of Martin Shreiner. He first appears in the city tax records of 1830. He and his older brother Martin, Jr. took over the clockmaking business of their father when he turned to making fire engines. Their clocks carry the signature "M. & P. Shreiner" and, like those of their father, are numbered. In 1835 Philip was chosen as an engineer of the new American Fire Engine and Hose Company of Lancaster. That same year he married, and in 1838 he and his wife left Lancaster to settle in nearby Columbia. Eventually a son joined him in business. In 1857 he was elected to serve as a member of the Columbia school board. The directory of 1874 lists him as retired at his home at 15 North Front Street. He is buried with his wife and children in the cemetery founded by his father in Lancaster.

Smith, Alexander Landry. *Strasburg*. (d. ca. 1819)

Smith first appears in the Leacock Township tax lists of 1796 as Alexander Londresmith. In 1799 he was listed as Alexander Landry Smith and from 1801 through 1816 as Landry Smith. In 1817 the name had reappeared in the "L" section as Landy Smith, but with no Christian name listed. He apparently died in 1819 because the entry for that year includes the word "Estate." Smith owned six acres of choice land in Strasburg and his shop was located on Main Street. No known clocks by him have survived.

Smith, Daniel. *Lancaster*.

Smith was taxed as a clockmaker from 1821 through 1824. In 1825 his trade was changed to "printer." His relationship to the Lancaster clockmaker John

[6]

CHEAPEST AND BEST CLOCK AND
WATCH ESTABLISHMENT IN THE CITY!!

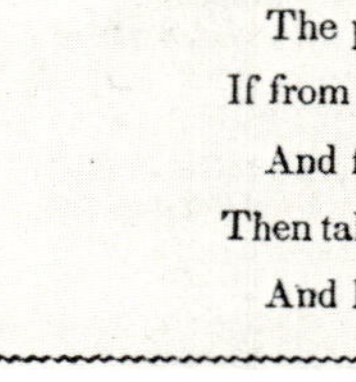

M. SHREINER,
NO. 26, NORTH QUEEN STREET.

Wind me with care and use me well
And let me have fair play,
And I, to you will try to tell
The precise time of day.
If from some cause I chance to stop
And fail to give the hour,
Then take me back to Shreiner's shop
And he will give me power.

1-27. Martin Shreiner (Jr.) ad, *J. H. Bryson's Lancaster Directory for 1843.*

Smith is unknown. No clocks by him have been located.

Smith, George. *Strasburg.*

George Smith is listed as a watchmaker in the Strasburg Township tax records from 1814 through 1825. In 1828 his trade was given as innkeeper. He owned property on the north side of Main Street. At least one clock bearing his name on the dial exists.

Smith, John. *Lancaster.*

First noted in 1817, Smith appears in the tax records through 1829. He put his name on at least one eight-day tall clock which has sweep calendar and seconds hands.

Smith, Joseph.

Joseph Smith is listed as a clockmaker in Earl Township from 1819 through 1822. None of his clocks have been located.

Stake, Christian. (d. 1799)

Stake owned property in Manor Township that he had purchased from James Logan. An advertisement placed in the *Lancaster Journal* of December 7, 1799 offered for sale "a complete sett of Clock & Watch-maker's tools" and a house clock from the estate of "Christian Stake, Esq., deceased." Farm utensils were also included. No clocks bearing his name have been located.

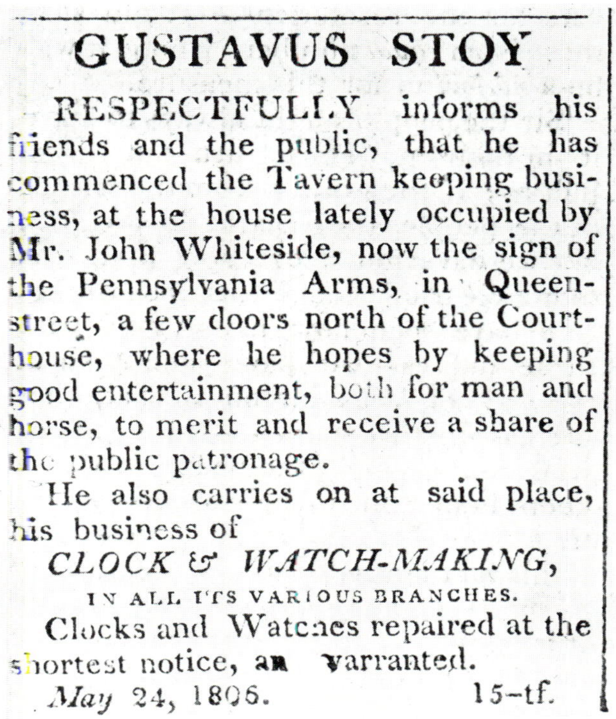

GUSTAVUS STOY

RESPECTFULLY informs his friends and the public, that he has commenced the Tavern keeping business, at the house lately occupied by Mr. John Whiteside, now the sign of the Pennsylvania Arms, in Queen-street, a few doors north of the Court-house, where he hopes by keeping good entertainment, both for man and horse, to merit and receive a share of the public patronage.

He also carries on at said place, his business of

CLOCK & WATCH-MAKING,

IN ALL ITS VARIOUS BRANCHES.

Clocks and Watches repaired at the shortest notice, and warranted.

May 24, 1806. 15–tf.

1-28. Gustavus Stoy ad, *Lancaster Journal*, May 24, 1806.

Stauffer, Samuel C. *Manheim.* (b. 1757, d. 1825)

Stauffer appears in the tax records of the town from 1783 through 1810. He made his clocks in a log building on Prussian (now Main) Street at the corner of Market Square. The site is now occupied by the Manheim National Bank. He is said to have been in partnership with fellow clockmaker Christian Eby. His known clocks have brass dials and are signed "Samuel Stauffer/Manhm." As they are numbered, it is known that he made at least thirty-four. Number 25 had a three-train quarter-hour chiming movement with gathering pallet locking.

Stehman, Henry. *Lancaster.*

Stehman started his business in 1829 and is noted through 1832. No known clocks by him survive.

Stoner, Rudy. *Lancaster.* (b. Jun. 25, 1728, d. Apr. 25, 1769)

Stoner is the first recorded clockmaker of Lancaster County. He appears in the tax records of 1754. The notes of the county commissioners show that he had charge of the tower clock of the first courthouse (which he may have made) through 1757, when Abrum Laroy was given the job. Stoner resumed care of the clock from 1765 until his death four years later. His shop and home were located just north of the courthouse, bordering on the market, immediately next door to the Borough's famous gunsmith and patriot, William Henry. The detailed inventory taken at Stoner's death, in addition to a complete set of clock and watch-maker's tools, parts, and patterns, includes cutting engines and fusee engines for both watch and clock work indicating that he may have been one of the few American craftsmen making watches in entirety.[2] Stoner was one of the charter members of Pennsylvania's third subscription library, the Juliana Library Company. He was also active in the Union Fire Company, and was a member of the Moravian church in which he was known as Rudolf Steiner. Only three thirty-hour clocks have been located, although it is believed that at least one eight-day clock is still in the county.

Stoy, Gustavus. *Lancaster.* (b. ca. 1777, d. May 25, 1816)

A son of William Stoy, pastor of the German Reformed Church of Lancaster from 1758 to 1763, and Maria Elizabeth Maus of Philadelphia, Stoy was probably born in the town of Lebanon where his father practiced medicine starting in 1767. His maternal grandfather, Frederick Maus, was a Philadelphia stocking weaver. The relationship with the Philadelphia clockmaker of the same name is unknown. Stoy made clocks in Lebanon from 1795 until 1805 when he moved

2. For the complete inventory see Stacy B. C. Wood, Jr., "Rudy Stoner, 1728–1769 Early Lancaster, Pennsylvania Clockmaker," *Journal of the Lancaster County Historical Society* 80, no. 2 (Easter 1976): 122–5.

his family to Lancaster Borough. There he became innkeeper of the Pennsylvania Arms on North Queen Street, a few doors above the courthouse. It was at the inn that he carried on his business of clock and watchmaking. He served as a private in the rifle regiment, 2nd Brigade of the Pennsylvania Militia in 1814 when it deployed to Baltimore. The inventory of his estate taken at the time of his death includes "1 Sett watchmakers tools" valued at $150.00, approximately one-sixth of the estate's total worth.

Witwer, Abel. *Ephrata.* (b. Oct, 2, 1767, d. Sept. 5, 1821)

Witwer appears in the Cocalico tax records of 1796 through 1798. He was a trustee of the Ephrata Cloister of the Pious and Ascetic Seventh-Day Baptists from 1815 to 1820. He is believed to have carried on the trade of clockmaker in the "house in the Meadow" (now the print shop) while he was one of the two remaining Brothers.

1-29. Abel Witwer Clock Shop, Pennsylvania Historical and Museum Commission, Ephrata Cloister.

Zahm, Godfried M. *Lancaster.* (b. Feb. 4, 1817, d. Mar. 14, 1895)

Zahm had been apprenticed to Martin (Jr.) and Philip Shreiner in 1832. In March of 1838 he opened a store in Columbia, but he returned to Lancaster in May of the same year and opened a store at 11 North Queen Street. There he made clocks, repaired watches, music boxes, accordions, and jewelry. He was also a silversmith. In 1844 he moved to the southwest corner of the Center Square. He owned a home at 53 East Chestnut Street and finally at 445 West Chestnut Street. He was largely responsible for the extension of the city to the northwest, where he ran in the gas and water lines at his own expense. At least two clocks are still extant. One is a regulator that was made for the watch factory, and the other is his Centennial musical calendar clock. He is said to have made five watches.

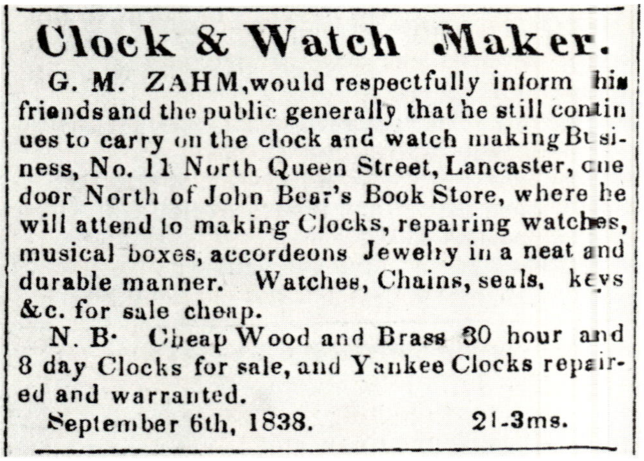

Clock & Watch Maker.

G. M. ZAHM, would respectfully inform his friends and the public generally that he still continues to carry on the clock and watch making Business, No. 11 North Queen Street, Lancaster, one door North of John Bear's Book Store, where he will attend to making Clocks, repairing watches, musical boxes, accordeons Jewelry in a neat and durable manner. Watches, Chains, seals, keys &c. for sale cheap.

N. B. Cheap Wood and Brass 30 hour and 8 day Clocks for sale, and Yankee Clocks repaired and warranted.

September 6th, 1838. 21-3ms.

1-30. Godfried M. Zahm ad, *Lancaster Examiner and Herald*, Sept. 6, 1838.

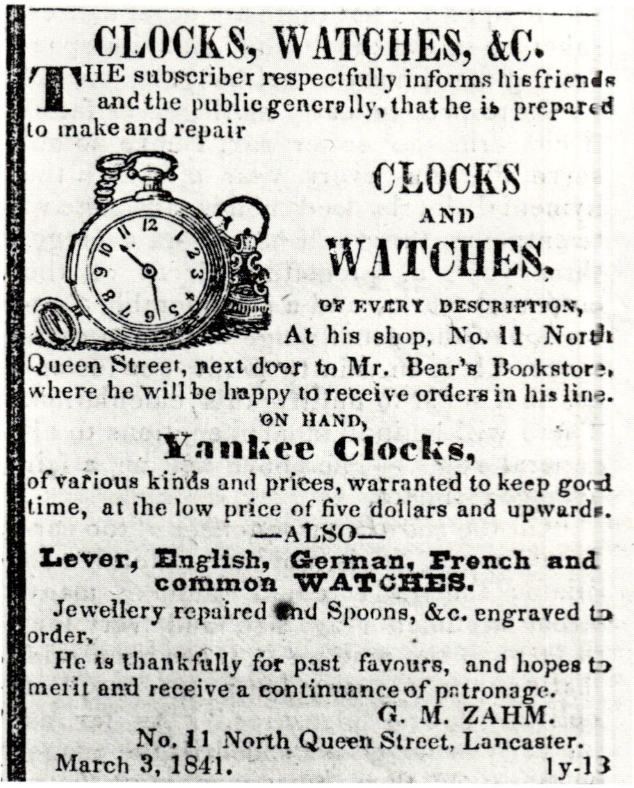

CLOCKS, WATCHES, &C.

THE subscriber respectfully informs his friends and the public generally, that he is prepared to make and repair

CLOCKS

AND

WATCHES,

OF EVERY DESCRIPTION,

At his shop, No. 11 North Queen Street, next door to Mr. Bear's Bookstore, where he will be happy to receive orders in his line.

ON HAND,

Yankee Clocks,

of various kinds and prices, warranted to keep good time, at the low price of five dollars and upwards.

—ALSO—

Lever, English, German, French and common WATCHES.

Jewellery repaired and Spoons, &c. engraved to order.

He is thankfully for past favours, and hopes to merit and receive a continuance of patronage.

G. M. ZAHM.

No. 11 North Queen Street, Lancaster.

March 3, 1841. 1y-13

1-31. Godfried M. Zahm ad, *Lancaster Examiner and Herald*, Mar. 3, 1841.

II CASES

For many reasons the cases of clocks from Lancaster County merit special attention. In general, it is the clock case, and not the movement or dial, that first attracts the viewer's notice. Indeed, it is likely that the original owners of the clocks illustrated in this book often esteemed a tall clock as their most important and expensive piece of furniture. Today, a collector often buys a clock on the basis of the appearance and quality of the case, while he may know very little about the quality or authenticity of its working parts.

These cases constitute a large and important body of Lancaster cabinetwork dating from the mid-eighteenth century through the third quarter of the nineteenth century. Hence, they provide a sound basis for defining many of the characteristics of regional cabinetmaking, and illustrate both the quality and the diversity of Lancaster County furniture.

The cases themselves are the primary source of information. Unfortunately, no case illustrated in this book is signed or labeled by its cabinetmaker. Further, very few clocks now have complete and accurate accounts of their early or original ownership. The extant account books of several joiners and cabinetmakers from Lancaster County provide some information, but too few documents of this nature survive. Consequently, our present knowledge of clock cases believed to have been made in Lancaster County must be based upon examination and analysis of the cases themselves as seen in the context of documentary evidence about the broader field of Lancaster County furniture. Every caution has been taken to ascertain that the cases and the movements shown here commenced their existence together. However, it should be stated that a number of "married" tall case clocks from Lancaster County do exist; in these instances, the works and cases were united at a time after the production of the respective components.

One basic assumption is that the movements and the cases generally were fabricated by different craftsmen. In other words, the clock works were made by a clockmaker, while the case was constructed by a cabinetmaker. There is very little, if any, evidence that any craftsman in the Lancaster area was both cabinetmaker and clockmaker. The woodworking artisans who fabricated cases were called either joiners or cabinetmakers; often, these terms were use interchangeably. The joiner was a person who could execute both architectural woodwork and furniture, whereas the cabinetmaker was sometimes more specialized.

Most of the cases illustrated here house movements for thirty-hour or eight-day clocks. They are now known as "grandfather" clocks, although this designation was not in use before the last quarter of the nineteenth century. Through the eighteenth and early nineteenth centuries, clock cases were usually designated according to the nature of their works and the wood; occasionally, the term "house clock" was employed. The basic form for the tall case clock developed in Europe in the mid-seventeenth century; its shape was dictated by functional requirements and architectural theory. The movement was housed in the hood section, while the waist section provided space for the weights and pendulum. The base section gave stability to an often top-heavy form. However, the theory of the column as the fundamental element in design and proportion in architecture and the decorative arts also influenced the basic design of the tall case clock. The hood corresponds to the rôle of the column's capital, the waist section approximates the column's shaft, and the bases of both columns and clock cases have many features in common. Within this basic formula of design, the ornament, treatment of the pediment (if present), and the detailing were influenced by prevailing styles in furniture and architecture.

Due to its ethnic diversity, Lancaster County was a meeting place for many styles in art throughout the century following 1750. Lingering Germanic sensitivities often mingled with English-oriented styles in a surprising and individualistic manner. However, the strongest stylistic influence on Lancaster County clock cases made in this period was the high-style English taste in furniture as manifested in Philadelphia cabinetwork. Since a clock case usually was an expensive piece of furniture, it reflected the most stylish and timely aspects of local taste. Only rarely could a clock case from Lancaster County be considered as a piece of folk art. Only a handful of clock cases partake in the tradition-oriented vocabulary of Germanic ornament which characterizes the noted chests and schranks with painted or inlaid decoration, and therefore, no examples of painted or wax-inlaid clock cases are shown here. The fact that the majority of the clock cases reflect English furniture styles is a telling indicator of the gradual acculturation of the Germanic elements which constituted the majority of Lancaster County's population. Moreover, it stands to reason that regional tastes in furniture were established by prevailing styles in the town of Lancaster, the countyseat which was founded in 1730. As the social, governmental, financial, and commercial center of a large county, it is only natural to conclude that the town, with a concentration of artisans, would exert a strong influence over the smaller communities and the surrounding countryside. It should be remembered that in the late eighteenth century Lancaster was America's largest inland town. It chronological sequence, the styles which flourished in the town provide a logical method of organizing the cases. These styles include Queen Anne, Chippendale, transitional from Chippendale to Federal, early Federal, late Federal, and Victorian.

The Queen Anne style flourished in the colonies from the 1730s to about 1760. In Philadelphia, the style was characterized by a sophisticated and graceful use of the S or cyma curve; it determined the outline of forms ranging from the cabriole leg to the scroll pediment. In Lancaster, the local interpretations of the Queen Anne style were simplified and somewhat provincial. The earliest manifestations of this style may be noted in the 1750s, and vestigial elements of the style sometimes lingered on at least as late as the end of the eighteenth century. The walnut case (Fig. 2-1) of the thirty-hour clock by Rudy Stoner of Lancaster, dating about 1754 to 1769, illustrates a simple, but effective development of the local Queen Anne style. The arched hood and shaping of the pendulum door employ the semicircular form favored in Queen Anne designs. This arched hood is a very rare feature in Lancaster County clock cases. A more elaborate version of a Queen Anne case is that housing works by Rudy Stoner illustrated in Fig. 2-9. The raised shell which graces the center of the tympanum of the pediment may be compared to the carved shells on the crest rails of Queen Anne chairs from the Philadelphia area. The effect of the tympanum filling the entire space between the scrolls of the pediment and the inverted vasiform turnings of the finials suggest that this design may be derived in part from German baroque cabinetwork. In this context, it should be noted that several joiners working in Lancaster between 1750 and 1770, including Philip Thomas, Johannes Grosch, and George Burkhart, would have been exposed to the German baroque style in their native land.[1]

The mainstream of the development of the Queen Anne style in Lancaster County clock cases from the second half of the eighteenth century is expressed in the numerous flat top pieces. Indeed, some of these cases are so simple that they almost defy any specific stylistic designation. In some, the presence of ogee bracket feet and fluted quarter columns betrays some degree of indebtedness to the prevailing Chippendale style. The majority of the flat-top cases are characterized by molded bases, slender waist sections with arched pendulum doors, and bold cove moldings crowning the hoods (Figs. 2-2 through 2-8). The base and waist sections of many of the flat top cases are ornamented with unfluted quarter columns in the front corners. The design of these unfluted or plain quarter columns, favored in the furniture of Chester and Lancaster Counties, is derived from the Tuscan Order in architecture. It is reasonable to believe that the many flat-top cases of this type housing works by John George Hoff were most likely made by joiners working in the town. The form was produced in rural areas of the county, as illustrated by the painted, simple pine case housing a movement by John Belsner of Ephrata (Fig. 2-5) and the walnut case housing works by Jacob Gorgas of Hinkletown (Fig. 2-8); the latter could date as late as 1810. The best of the flat-top cases epitomize the slender, linear grace with minimal ornamentation associated with the Queen Anne style.

Lancaster County's achievements in furniture have long received recognition for the quality of the regional Chippendale school. The development of Lancaster's Chippendale school was obscured for many years by the unsubstantiated practice of attributing all richly carved Chippendale pieces to John Bachman II; these attributions were based upon incorrect and incomplete evidence published in two articles.[2] Recent research, published in 1974, indicates that John Bachman II (1746–1829) did not produce elaborate carved furniture.[3] In fact, most of his work was of a substantial but very plain nature, and it is likely that the majority of the clock cases listed in his extant account book were simple, flat-top pieces; it is significant that the most expensive case ever produced by his workshop cost £10 in 1790. This price probably indicates a good quality flat-top case.[4]

The Chippendale style in Lancaster started in the mid 1760s and remained popular at least through the first decade of the nineteenth century. This style pays homage to Thomas Chippendale, a London cabinetmaker who published his *The Gentleman and Cabinet-maker's Director* in 1754. This book of designs for

1. John J. Snyder, Jr., "Chippendale Furniture of Lancaster County, Pennsylvania 1760–1810" (MA thesis, Winterthur Program, University of Delaware, 1976).

2. Carl W. Drepperd, "Furniture Masterpieces by Jacob Bachman," *American Collector* 14, no. 9 (October 1945): 6–20; Samuel E. Dyke, "The Bachman Family of Cabinetmakers 1766–1897," *Journal of the Lancaster County Historical Society* 69, no. 3 Trinity (1965): 168–80.

3. John J. Snyder, Jr., "The Bachman Attributions: A Reconsideration," *Antiques* 55, no. 5 (May 1974): 1056–65.

4. John J. Snyder, Jr., "Chippendale Furniture of Lancaster County, Pennsylvania 1760–1810" (MA thesis, Winterthur Program, University of Delaware, 1976).

furniture was of great importance in spreading the rococo style as applied to decorative arts. It was the third edition of the *Director*, published in 1762, however, which exerted definite influence in the Philadelphia area. Of great importance for introducing Chippendale designs to Lancaster was the presence of Philadelphia-made furniture in the homes of the town's elite. There is strong circumstantial evidence that George Burkhart (1721–1783), Lancaster's richest joiner before the American Revolution, had access to a copy of the 1762 printing of the *Director*. The Lancaster Chippendale school developed between 1760 and 1810 through a blending of Philadelphia interpretations of the rococo with older traditions of German baroque. In the half century following 1760, more than one-hundred-and-sixty woodworking craftsmen worked in the town of Lancaster alone. The outstanding characteristic of Lancaster Chippendale furniture is the extensive use of relief carving; this differs from the customary practice in Philadelphia, where carving was usually applied.

The earliest documented piece of the Lancaster Chippendale school is the case of the organ at Lancaster's First Reformed Church, made by George Burkhart between 1769 and 1770.[5] However, the very handsome cherry case of the clock housing works by Wilton Atkinson of Lancaster, dating about 1770 to 1773 (Fig. 2-10) shows that the new style was making its impact on domestic furniture at the same time. The detailing of the carved floral rosettes which terminate the scrolls of the pediment, and the shape of the flame finials, are characteristic of the best attributes of this style. A very unusual feature in Lancaster cabinetwork is the open fretwork which fills the tympanum of the scroll pediment of a case housing a movement by George Hoff (Sr.) (Fig. 2-11). Certainly more representative of the simple elegance which characterizes many Lancaster Chippendale clock cases is the case of the clock by Martin Shreiner illustrated in Fig. 2-12. Several cases which may be associated with the works of joiners in Lancaster during the last two decades of the eighteenth century employ a restrained use of applied carving, often of a foliate nature, in the central area of the pediments (Figs. 2-13 and 2-14).

The masterpieces of Lancaster Chippendale clock cases are the examples with relief carving in the pediments; fewer than a dozen authentic examples are known. Although none of these cases is documented to the maker, the best evidence supports the belief that most were made in the town, although a few may have been made in communities like Strasburg or Manheim (Figs 2-15 and 2-16). The tympanum of the pediment is filled with an undulating, but symmetrical design of vine and foliate motifs, all carved from solid wood. In both technique and overall design, the carving reveals a marked indebtedness to German baroque ornament (Fig. 2-58).

It is possible that even these elaborate cases do not represent the greatest accomplishments of Lancaster's cabinetmakers in the last third of the eighteenth century. Even the greatest relief-carved clock cases do not approach the sophistication of local organ cases made in this period.[6] Further, the reference to "Mahogini boards" (sic) in the 1783 estate inventory of George Burkhart implies that there may be some elaborate Lancaster clock cases in mahogany that are now mistaken for Philadelphia work.

The case of the clock with works by Jacob Gorgas Jr. probably dating about 1780 to 1790 (Fig. 2-17) illustrates the strength of the style centered in the town as it exerted an influence on a joiner working in a rural context, here, probably near Hinkletown. This case combines a delightful naive quality with boldness in proportion and execution. A number of accomplished Chippendale cases with works by Manheim makers testify to that town's role as a secondary center for cabinetmaking. Indeed, there is some evidence that a distinct, subsidiary school of Lancaster relief carving flourished in Manheim and the immediate environs. Although no example of a fully relief-carved clock case from Manheim is shown here, the illustrated examples (Fig. 2-18 through 2-21) reveal something of the diversity of cabinetwork in one small region; these four cases most likely represent the work of three different shops. The differences in carving the typical Lancaster County twelve-pointed floral rosette should be noted.

An interesting walnut Chippendale case with unsigned works (Fig. 2-22) has been attributed by tradition to Musser John Landis (1761–1827). However, there is no documentation that Musser John Landis was a joiner and cabinetmaker, and in tax records he appears as a farmer. Thus it seems probable that this clock was originally owned by Musser John Landis, who lived on a farm between Lancaster and Lampeter. The carving of the rosettes is an unusual variant.

Although neoclassicism from England was introduced to the Philadelphia area on the eve of the Revolution, it made little impact in general on furniture designs until the following decade. The most influential printed sources for neoclassicism as it was manifested in the Federal style in America were George Hepplewhite's *Cabinet Maker and Upholsterer's Guide* of 1788 and Thomas Sheraton's *Cabinet-Maker's and Upholsterer's Drawing-Book*, published in sections from 1791 to 1794. Nonetheless, the appearance of these two books did not mark an abrupt and total change to the new Federal style in Lancaster County cabinetwork. Many clock cases made in the two decades following 1785 possess stylistic features which mark them as being transitional from the Chippendale to Federal styles.

5. John J. Snyder, Jr., "Carved Chippendale Case Furniture from Lancaster, Pennsylvania," *Antiques* 57, no. 5 (May 1975): 964–75.

6. Ibid., p. 965.

An excellent example of this transitional stage is seen in the elegant walnut case housing a movement by John Eberman, Jr. of Lancaster, probably dating about 1785–1795 (Fig. 2-24). The fluted quarter columns and basic proportions bespeak a lingering influence of the Chippendale style. However, the reeded frieze applied at the top of the waist section, the outflaring feet with scalloped skirt, and the delicate scroll pediment terminated with small turned roundels as rosettes reveal the new sensitivity to light, linear ornament. The use of figured walnut also introduces a note of subtle color, heralding the coloristic interest later to be expressed through veneers and inlays. The case of a clock housing a movement by John Hoff of Lancaster, probably dating between 1800 and 1810, employs light colored inlays for the inverted shell motif beneath the central finial and also for the rosettes which terminate the scrolls of the pediment (Fig. 2-25). The feet, fluted quarter columns in base and waist sections, and cursive outline of the top of the pendulum door remain in the Chippendale style.

Cases from many parts of the county indicate that the period between 1795 and 1810 was often characterized by this transitional style in cabinetwork. The fact that the face of a clock with movement by John Leinbach of Reamstown is dated 1797 establishes a firm point of reference in this transitional stage (Fig. 2-26). This case may be representative of the products of a now unidentified workshop in the Ephrata/Reamstown area which very likely produced cases for clocks by Solomon Gorgas and Abel Witwer illustrated in Figs. 2-27 and 2-28.

The vine and berry inlay in the tympanum of the pediment of the case housing works by George Fisher of Lancaster (Fig. 2-29), dating between 1799 and 1808, is very unusual in Lancaster County cabinetwork. It may be related to somewhat similar inlays frequently associated with Chester County.[7] The treatment of the quarter columns with fine horizontal ribbing appears to be related to similar ornament found on woodwork in houses near Terre Hill in eastern Lancaster County, and it is plausible that this case may have been made in that region. Certainly this case bears a distinct relationship to the inlay which graces the case with works by John Davis of New Holland (Fig. 2-46).

The case with works by Christian Hall of Lititz (Fig. 2-30) demonstrates the tenacity of the flat top form here united with new styles of ornament. The contrasting veneers, and particularly the crossbanded borders of the waist and pendulum door, may reveal Baltimore influence.

The fully developed Federal style was characterized by the extensive use of inlays and veneers, attenuated proportions, oval forms, and an overall feeling of delicacy and lightness. Six clock cases may show the main characteristics of early Federal clock cases from the town of Lancaster (Figs. 2-31 through 2-36). Favored treatments of the rosettes terminating the scrolls of pediments include inlaid light wood paterae (Fig. 2-31) or turned roundels (Fig. 2-33); this latter device may be derived from contemporary architectural ornament. The typanum of the pediment is sometimes veneered with figured wood with the grain running vertically (Fig. 2-32), or it may be highlighted with matched veneers with the grain running on the diagonal from the median (Fig. 2-33). Side windows in the hood, in addition to the usual rectangular form, may sometimes be oval (Fig. 2-35) or lancet-shaped (Fig. 2-34). Frequently the columns of the hood are inlaid with light stripes and mounted in brass bases and capitals.

The account book of the Lancaster clockmaker John Hoff, (now in the collection of the Lancaster County Historical Society) covering the years from 1800 to 1816, indicates that most of his clients obtained their clock cases from cabinetmakers of their choice. Twice Hoff mentioned that he had delivered movements to an otherwise unidentified cabinetmaker named "Lichtendaler," and in 1807 he exchanged one movement for two clock cases from a Columbia cabinetmaker, William Green. In June, 1802 Hoff obtained a clock case valued at £9 15s from the Lancaster cabinetmaker, Christian Rine. It thus seems plausible that Rine may have been one of the cabinetmakers who produced some of the early Federal period clock cases which may be associated with Lancaster. However, this account book shows that Hoff had customers in places as far removed as western Pennsylvania and North Carolina, so it stands to reason that some of his works were not housed in locally made cases.

A cabinetmaker who very likely could have made some of the early Federal period clock cases was Moravian, Michael Stoner (1764–1810), who appeared in Lancaster sometime between 1798 and 1800 and worked there until his death. Stoner was both a cabinetmaker and chairmaker, and he sometimes worked in partnership with his cousin, George Thomas. The following entries in Stoner's estate inventory (in the collection of Lancaster County Historical Society), taken by John Smith and John Sener on March 15, 1810, prove that his shop produced clock cases and documents the existence of the tools needed to do inlaying and veneering:

1 Fennier (sic) Saw & key	£1-10-0
1 Fennier Screw	£0- 2-6
1 Glew Pot	£0-18-9
1 Frame for a Fennir Saw	£0- 3-9
1 Fennier Bench	£0- 1-10½
1 new Clock case (Cherry)	£6- 0-0
1 old do. Walnut	£1-17-6

Clock cases from Manheim and the adjacent area rival, and even excel, the quality of the cases produced

7. Margaret Berwind Schiffer, *Furniture and Its Makers of Chester County, Pennsylvania* (Philadelphia: University of Pennsylvania Press, 1966), pp. 262–3.

in the town of Lancaster in the Federal period. The exact reason for this is not known, but Manheim's position on north-south routes of travel may have given it a degree of importance not indicated by its size alone. For three decades following 1790, Manheim boasted a large number of joiners and cabinetmakers, although the makers of its best Federal style cases are unknown. The cases housing works by John Heinselman and Christian Eby (Figs. 2-37 and 2-38) suggest both the quality and diversity of cabinetwork from Manheim. Yet the greatest achievement of Federal clock cases from Lancaster County are the elegant eagle inlaid cases which usually house works by Jacob Eby of Manheim (Figs. 2-39 and 2-40). Of these eagle inlaid cases, of which no more than twenty exist, the greatest is the richly inlaid mahogany bowfront case at the Henry Francis du Pont Winterthur Museum (Fig. 2-40). In both technique of construction and finesse of execution, this case is a masterpiece. The eagle inlays which grace the waist sections of these clocks (Fig. 2-59) are similar in general design, but vary in detail. Some are heightened with stripes of red wax, with " E PLURIBUS UNUM" incised on the banner waving from the eagle's beak. It seems likely that these eagle medallions were produced by a now unknown specialist who supplied them to several cabinetmakers.

Federal clock cases from other areas of Lancaster County sometimes could approximate the stylistic qualities of those produced in the town, as seen in the examples housing works by Joseph Gorgas of Ephrata and George Smith of Strasburg (Figs. 2-41 and 2-43). It is logical to believe that sophisticated cabinetwork was produced in Marietta and Columbia at the opening of the nineteenth century, and the cases housing works by Thomas Dickey of Marietta (Fig. 2-44) and John Esterle of Maytown (Fig. 2-45) may well represent products of now unknown workshops from these towns on the Susquehanna River. In particular, the case of the clock by John Esterle, dated 1816, boasts an exceptional use of figured mahogany veneers. On the other hand, some clock cases from the Federal period evoke earlier stylistic elements (Fig. 2-42), whereas a few simplify the architectural components of the pediment to an almost abstract degree, as seen in the case housing works by Samuel Breneisen of Adamstown (Fig. 2-47).

About 1820 a later phase of Federal clock cases developed throughout Lancaster County. These are characterized by small turned feet of varying design and detail, broader proportions of the waist section, and heavier moldings. There is almost invariably a frieze of figured veneer at the top of the waist section. This basic type of case was produced in the city of Lancaster and surrounding communities with remarkable consistency at least as late as the 1840s (Figs. 2-49 through 2-52). Occasionally some of these cases betray some detail of the Empire style, as seen in the applied half columns with a variant of the Ionic Order for capitals on the base of the case housing a movement by Henry Ober (Jr.) of Elizabethtown (Fig. 2-52).

There is convincing documentary evidence for attributing the cherry and mahogany veneer case housing works by A. W. Baldwin of Lampeter Square to the workshop of Jacob Bachman. (Fig. 2-50) This Jacob Bachman (1798–1869) was a grandson of John Bachman II; he worked as a cabinetmaker in Bachmantown, in East Lampeter Township, near Soudersburg. His extant day book, covering the years from 1822 to 1863 (now in the Downs Manuscript Library of the Henry du Pont Wintherthur Museum in Wintherthur, Delaware), records that he frequently supplied Anthony Baldwin with "2 Cherry Clockcases" valued at $19.00 each in exchange for one clock movement valued at $44.00. This provides a valid argument for attributing the cases of A. W. Baldwin clocks to the workshop of Jacob Bachman.

The Victorian period, spanning the last six decades of the nineteenth century, gradually witnessed the decline of a regional school of cabinetmaking in the face of competition from factory-made furniture. However, an occasional tall case clock of note was produced as late as the time of the centennial, as seen in the case of the Zahm clock (Fig. 2-54). An interesting case which might be considered as Victorian in spirit, if not entirely in date, is that which houses a movement dated 1834 by Samuel Breneisen of Reamstown (Fig. 2-53). The basic design of this case follows local tastes of the late Federal period, but the applied panels, the carved face (Fig. 2-60) beneath the center finial, and the motifs of the British Coat of Arms, apparently decals, are atypical of an 1834 date. These decorative motifs do accord with features of mid-Victorian furniture, and it is thus possible that this case was embellished in the third quarter of the nineteenth century.

In the first half of the nineteenth century Lancaster County occasionally produced some highly individualistic clock cases, such as the shelf or mantel clock with face signed by Samuel Grosch of Marietta, dating about 1820 (Fig. 2-55). There is a tradition that this mahogany case was made by a member of the Hiestand family in Marietta. Another very unusual regional development was the placement of a clock in the center of a corner cupboard, as seen in the example with works by John Hoff, made for Daniel Sittle of Lampeter Square in 1812 (Fig. 2-57). Fewer than six of these corner cupboard clocks are known to survive.

It is hoped that this discussion together with the following illustrations provides an accurate cross section of clock cases made in Lancaster County from the Queen Anne through Victorian styles. The descriptions of cases are based chiefly upon matters of style, for space precluded illustrating structural details. It should be noted that woods are identified according to visual examination, and not by microanalysis.

Fig. 2-1. Arched hood case housing works by Rudy Stoner of Lancaster, dating about 1754 to 1769. The arched hood, bold cove molding at the base, and beaded corners of the waist section bespeak the early date. Walnut primary wood, 83¼″ high. (Private collection)

Fig. 2-2. Flat-top case housing works by John George Hoff, Sr. of Lancaster, dating about 1769 to 1785. This is one of the finest flat-top clock cases from Lancaster County; it is distinguished by its molded platform on the hood, crisp cove molding of the cornice, and narrow waist section. Walnut primary wood, 85½″ high. (Collection Lancaster County Historical Society)

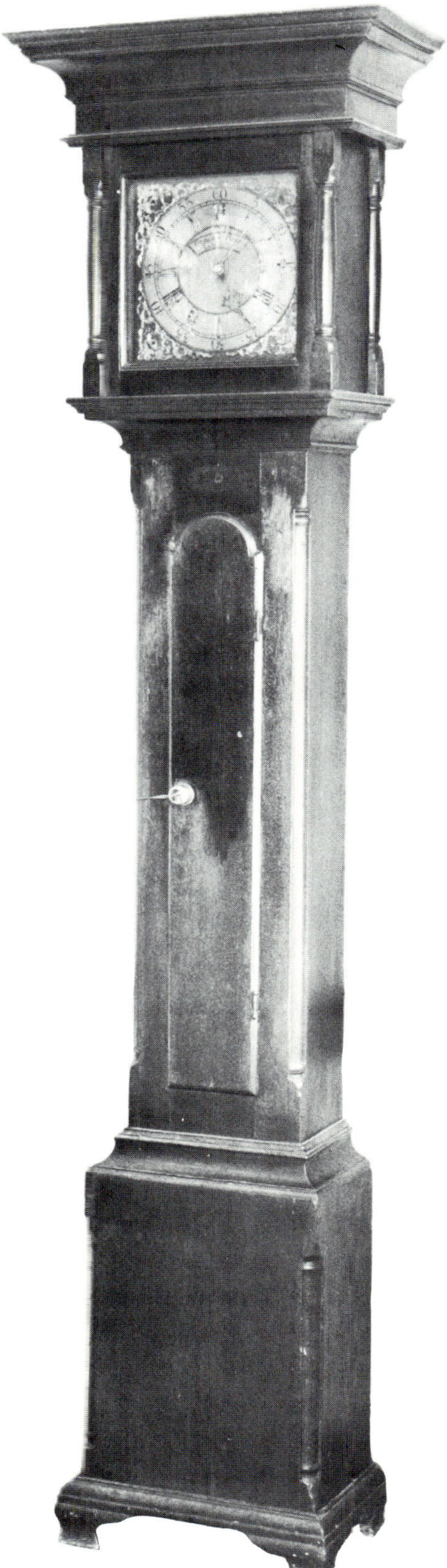

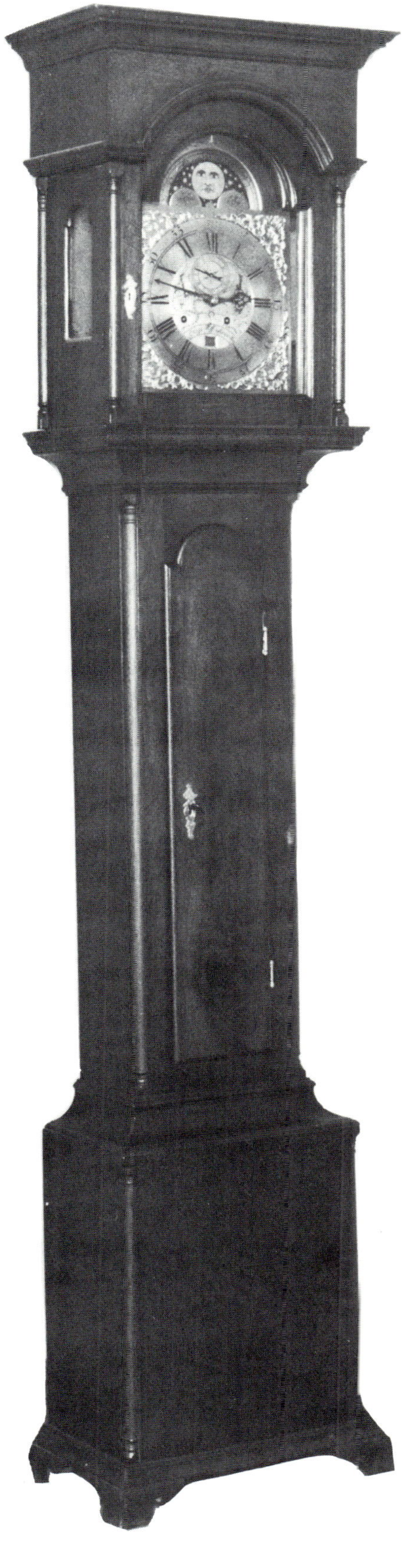

Fig. 2-3. Flat-top case housing thirty-hour movement by John George Hoff, Sr. of Lancaster, dating about 1769 to 1785. The ogee bracket feet, molded cornice details, and lambs' tongues which terminate the unfluted quarter columns are pleasing variations to the usual Lancaster flat-top clock case. Walnut primary wood, 90½″ high. (Collection of Pennsylvania Farm Museum of Landis Valley)

Fig. 2-4. Flat-top case housing eight-day works by John George Hoff, Sr. of Lancaster, dating about 1769 to 1785. This is a well-proportioned example of the basic form favored in Lancaster County flat-top Chippendale cases. Note the arched windows on the sides of the hood. Walnut primary wood, 86″ high. (Private collection)

Fig. 2-5. Flat-top case with eight-day works by John Belsner of Ephrata, predating 1785. This pine case bears an old, painted surface which may be representative of the finishes favored for soft-wood clock cases in eighteenth century Lancaster County. This case, made without quarter columns in the base and waist, and lacking columns or side windows in the hood, represents a simple, rural development of the form. Pine primary wood, 84¾″ high. (Private collection)

Fig. 2-6. Flat-top case housing thirty-hour movement by Martin Shreiner of Lancaster, dating about 1795 to 1805. The evidence for dating this clock, based on Shreiner's practice of numbering works, indicates the great tenacity of the flat top clock case. The fluted quarter columns and the curved top of the pendulum door are pleasing refinements. Walnut primary wood, 87½″ high. (Collection of Pennsylvania Farm Museum of Landis Valley)

Fig. 2-7. Flat-top case housing thirty-hour works by John Eberman, Jr., of Lancaster, dating about 1785 to 1807. This simple Chippendale case illustrates the retention of older styles, even in Lancaster borough, after the Revolution. Walnut primary wood, 85¾″ high. (Private collection)

Fig. 2-8. Flat-top case housing thirty-hour movement by Jacob Gorgas Jr. of Hinkletown, dating about 1791 to 1810. This is a simple version of the customary design for a flat-top clock case, fabricated without quarter columns. The vertical stiles of the front of the waist section are pegged into the side boards. Walnut primary wood, 87¼″ high. (Private collection)

Fig. 2-9. Scroll-pediment case housing thirty-hour movement by Rudy Stoner of Lancaster, dating about 1754 to 1769. The raised, carved shell which fills the central area of the tympanum of the pediment is a very unusual device; this is one of the few Lancaster clock cases with pronounced Queen Anne characteristics. The central finial with its supporting fluted plinth is shown beside the clock. It is possible that the straight bracket feet have lost some height. Walnut primary wood, 92″ high. (Private collection)

Fig. 2-10. Scroll-pediment case housing eight-day movement by Wilton Atkinson of Lancaster, dating about 1770 to 1773. The brief working period of this maker provides an accurate basis for considering this case as being one of the earliest examples of the Chippendale style in Lancaster County. The bold ogee bracket feet, the raised, scalloped panel on the base, the carved floral rosettes which terminate the scrolls of the pediment, and the flame finials are characteristic of the Lancaster Chippendale school. Cherry primary wood, 102¼″ high. (Private collection)

Fig. 2-11. Scroll-pediment case housing works by John George Hoff, Sr. of Lancaster, dating about 1769 to 1790. The design of the base and waist sections, with unfluted quarter columns, is typical of Lancaster cabinetwork of the last four decades of the eighteenth century. Very unusual features include the open fretwork filling the tympanum of the scroll pediment above the horizontal cornice molding and the motif of the floral rosettes. Also notable are the high, rectangular side windows on the hood, and the turnings of the hood's columns. Walnut primary wood, 93″ high. (Private collection)

Fig. 2-12. Scroll-pediment case housing eight-day movement by Martin Shreiner of Lancaster, dating about 1796. This is an excellent example of regional Chippendale cabinetwork, with fluted quarter columns, arched pendulum door, and typical Lancaster twelve-pointed floral rosettes terminating the scrolls of the pediment. The small scale of the urn-shaped finials may indicate some influence from Federal furniture styles. Walnut primary wood, 97¾″ high. (Private collection)

Fig. 2-13. Scroll-pediment Chippendale case housing eight-day works by John Eberman, Jr., of Lancaster, dating about 1785 to 1800. Notable features of this case include the applied, foliate carving in the center of the tympanum of the pediment, the richly detailed carving of the floral rosettes which terminate the scrolls of the pediment, and the flame finials which rise directly from the supporting plinth without an intermediate urn or bulbous section. This clock descended through many generations of the Evans family. Walnut primary wood, 99¼″ high. (Private collection)

Fig. 2-14. Scroll-pediment case housing eight-day works by John Gottlieb Eberman of Lancaster, dating about 1782 to 1801. The form of the unusual applied carving in the tympanum of the pediment complements the plumelike device which constitutes the center finial. The typical Lancaster Chippendale floral rosettes exhibit good carving. Walnut primary wood, 90¼″ high. (Private collection)

Fig. 2-15. Carved Chippendale scroll-pediment case housing eight-day movement by Christian Forrer of Lampeter. Although Christian Forrer worked from 1754 to 1774, this case could date between 1765 and 1795. It exhibits the best characteristics of elaborate relief carving, which is generally regarded as being the greatest achievement of the Lancaster Chippendale school. The elaborate relief carving filling the entire tympanum of the pediment, the rich flame finials, the large floral rosettes, and the carved molding of the scroll pediment are outstanding. Small details worthy of note are the arched doors on the sides of the hood and the incised lambs' tongues which terminate the fluted quarter columns in the base and waist sections. Walnut primary wood, 98″ high. (Collection the Virginia Museum of Fine Arts, Richmond, Virginia)

Fig. 2-16. Scroll-pediment case housing eight-day movement by Thomas Bradin Burrowes of Strasburg, dating about 1787 to 1810. This is one of the best proportioned examples of a Lancaster County relief carved Chippendale clock case. The cursive, foliate relief carving in the tympanum of the pediment is of high quality. The applied leaf motifs in the corners of the raised, scalloped base panel are an unusual feature. The detail of the three carved moldings is also notable, as are the front columns of the hood, which match the treatment of the fluted quarter columns with incised terminals in the waist and base sections. Before 1880 this clock belonged to the Lantz family living near Strasburg. Walnut primary wood, 105″ high. (Private collection)

Fig. 2-17. Scroll-pediment case housing eight-day works by Jacob Gorgas, Jr., near Hinkletown. This interesting case probably dates about 1780 to 1790. The strong moldings, raised base panel with hollow corners, and mitered construction of the front of the base section are unusual features. The large floral rosettes which terminate the scrolls of the pediment, and the rather primitive carved device below the plinth of the central finial, bespeak the hand of a rural joiner who was influenced by the styles flourishing in Lancaster borough. Walnut primary wood, 96¼″ high. The feet and finials are restorations. (Private collection)

Fig. 2-18. Scroll-pediment case housing eight-day movement by Samuel C. Stauffer of Manheim, dating about 1792 to 1807. The sunken, octagonal panel in the base, the geometric form of the rosettes which terminate the scrolls of the pediment, and the distinctive shape of the graceful urn finials have been noted in one other Chippendale clock case from Manheim. An unusual stylistic feature is the combination of fluted quarter columns in the base section with unfluted quarter columns in the waist section. The ogee feet have apparently lost some of their original height. Walnut primary wood, 91¾″ high. (Hershey Museum of American Life)

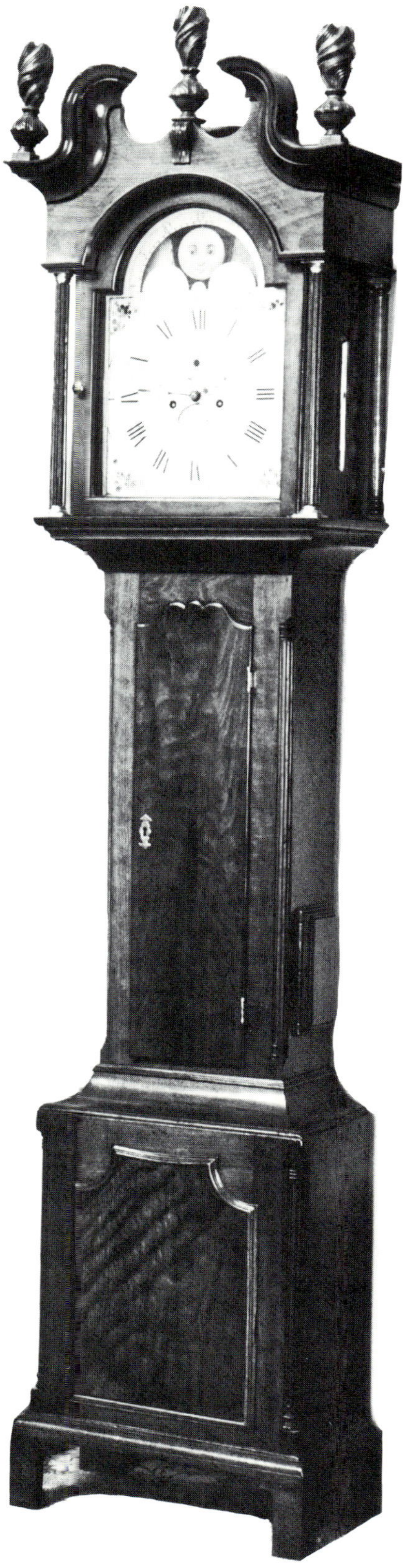

Fig. 2-19. Scroll-pediment case housing eight-day works by John Conrad Heinselman of Manheim, dating about 1792 to 1803. The high bracket feet, applied volute or console bracket below the center finial, and excellent flame finials are notable features. Unusual elements are the side "pockets" with molded covers on the waist section, and the manner of terminating the scroll pediment without rosettes. Walnut primary wood, 95½" high. (Hershey Museum of American Life)

Fig. 2-20. Scroll-pediment Chippendale clock case housing movement by John Conrad Heinselman of Manheim, dating about 1792 to 1803. The shaping of the base panel with upper hollow corners, the form of the pendulum door, the general design of the hood, and the detailing of the flame finials and the central console bracket suggest that this fine case and the preceding example may have been made by the same craftsman in the Manheim area. Walnut primary wood, 102½" high. (Private collection)

Fig. 2-21. Scroll-pediment Chippendale case housing eight-day works by John Conrad Heinselman of Manheim, dating about 1792 to 1803. This simple case probably represents the work of yet another unidentified joiner in the Manheim area. The rosettes which terminate the scrolls of the pediment are smaller than those encountered on most Lancaster County Chippendale clock cases. Cherry primary wood, 99½″ high. (Private collection)

Fig. 2-22. Scroll-pediment Chippendale case with unsigned works, probably dating about 1780 to 1805. This case exhibits many characteristics of local Chippendale cabinetwork, although the manner of carving the rosettes which terminate the scrolls of the pediment is unusual. Walnut primary wood, 95¼″ high. (Lancaster Mennonite Conference Historical Society)

Fig. 2-23. Scroll-pediment case housing eight-day works by Jacob Gorgas (Sr.) of Ephrata, dated 1771. The case, with molded rectangular base panel and molded frames on the side windows of the hood, appears to be contemporary with the date of the movement. The thin molding of the scroll pediment, and the roundels which form the rosettes are probably an updating of the case done in the Federal period, about 1790 to 1810. Walnut primary wood, 91½″ high. (Collection Lancaster County Historical Society, Lancaster; bequest of Mabel H. Nolt)

Fig. 2-24. Scroll-pediment case with eight-day movement by John Eberman, Jr. of Lancaster, probably dating about 1785 to 1795. This is an excellent example of the transition from the Chippendale to the Federal styles. Although the fluted quarter columns belong to the Chippendale style, the outflaring bracket feet, scrolled skirt, reeded frieze at the top of the waist section, and thin scroll pediment with turned roundels serving as rosettes bespeak definite Federal influence. The oval side windows on the hood also reveal the Federal period's interest in oval forms. The use of richly figured walnut is notable. Walnut primary wood, 94½″ high. (Private collection)

Fig. 2-25. Scroll-pediment case with thirty-hour works by John Hoff of Lancaster dating about 1799 to 1818. Although the bracket feet, fluted quarter columns, and shaping of the pendulum door are in the Chippendale style, the inlaid rosettes and inlaid shell-like device beneath the center finial mark this as a sophisticated transitional piece. Mahogany primary wood with light inlays, 94½″ high. (Private collection)

Fig. 2-26. Scroll-pediment case housing thirty-hour movement by John Leinbach of Reamstown; the painted face is dated 1797. The ogee feet and fluted quarter columns are in the Chippendale style; the simple molding of the scroll pediment with turned roundels as rosettes bespeaks the transition to the Federal style. Walnut primary wood, 95¼″ high. (Private collection)

Fig. 2-27. Scroll-pediment case with eight-day works by Solomon Gorgas of Ephrata, dating about 1787 to 1806. Another example of the transition from Chippendale to Federal styles, this case was possibly made by the same unknown joiner who produced the preceding example. This case very likely once had ogee bracket feet. Walnut primary wood, 94½" high. (Private collection)

Fig. 2-28. Scroll-pediment case housing thirty-hour works made by Abel Witwer of Ephrata, probably dating 1790 to 1800. This transitional style case may well represent the work of the maker of the preceding two examples. The combination of fluted quarter columns in the base section with chamfered corners in the waist section is unusual. The total design of the pediment is more fully developed than in the Leinbach clock illustrated in Fig. 2-26. Walnut primary wood, 94¾" high. (Private collection)

Fig. 2-29. Scroll-pediment case housing eight-day movement by George Fisher of Lancaster, dating about 1799 to 1808. This is a very unusual example of the transition from the Chippendale to the Federal styles. Whereas the floral rosettes which terminate the scrolls of the pediment are typical of Lancaster Chippendale cabinetwork, the raised base panel with bold, concave bevel, and the quarter columns with horizontal ribbing, are very unusual in this locality. The graceful and symmetrical vine-and-berry inlay in the tympanum of the pediment may be related to similar inlay associated with Chester County. Cherry case with inlays, 99″ high. (Amish Farm and House)

Fig. 2-30. Flat-top case housing eight-day works by Christian Hall of Lititz, dating about 1800. This unusual flat-top case retains Chippendale characteristics in its ogee bracket feet, the Gothic crest of the pendulum door, and the retained use of the flat-top form. Entirely in the Federal style is its ornament, including the reeded frieze at the top of the waist, the crossbanded borders of the base and pendulum door, and the waist's chamfered corners with inlaid stripes. In particular, the crossbanded borders might be indicative of Baltimore influence. Mahogany primary wood with veneers and inlays, 84″ high. (Private collection)

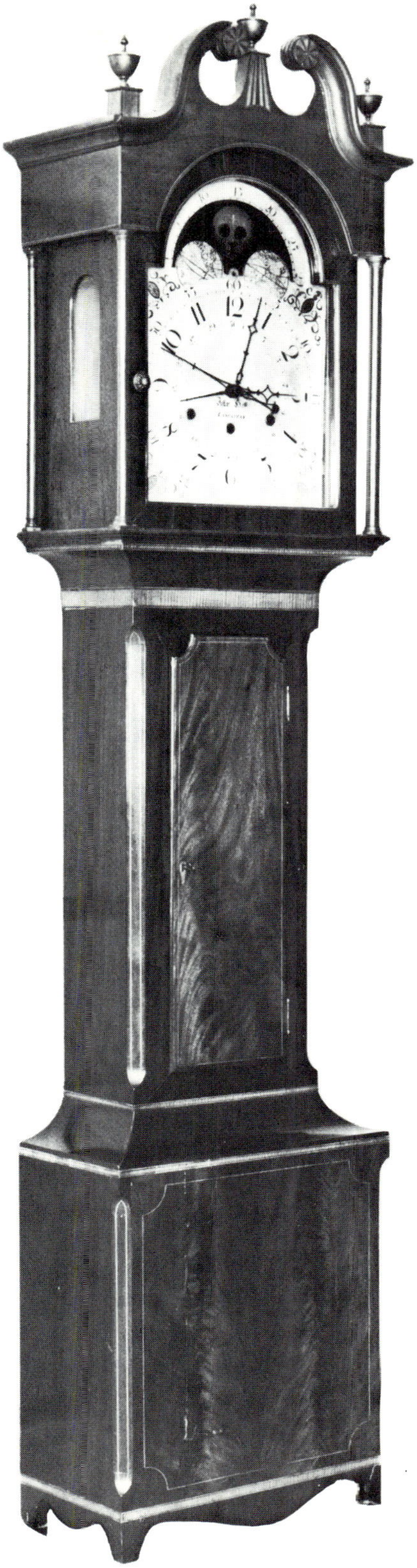

Fig. 2-31. Scroll-pediment case housing eight-day works which chime on the quarter-hour by John Hoff of Lancaster. The account book of John Hoff reveals that these works were made for John Kindig in 1810. The inlaid rosettes which terminate the scrolls of the pediment, and the inverted, shell-like motif beneath the central finial are characteristic of clock cases made in Lancaster borough in the Federal period. The applied frieze of curly maple at the top of the waist is an interesting and colorful feature. The columns at the hood, with inlaid light-wood stripes and brass bases and capitals are also encountered on many Lancaster County clock cases from the Federal period. The feet have possibly lost some of their original height. Mahogany primary wood with light-wood inlays, 93″ high. (Collection of Lancaster County Historical Society; gift in memory of Kendig H. Bare)

Fig. 2-32. Scroll-pediment case housing eight-day movement by John George Hoff of Lancaster, probably dating 1800 to 1810. It is not certain whether this movement was made by the father John George Hoff or his son. This case exhibits many of the characteristics of the best Federal-period cabinetwork associated with Lancaster borough, including the use of vertical panels of crossbanding as seen here on the base, and the wavy lines inlaid on the chamfered corners of the waist section. The practice of covering the tympanum of the pediment with vertically figured veneers is well illustrated here. Due to the fact that the turned feet probably were added at a later date, the original height of this clock cannot be ascertained. Mahogany primary wood with figured veneers and light inlays. (Private collection)

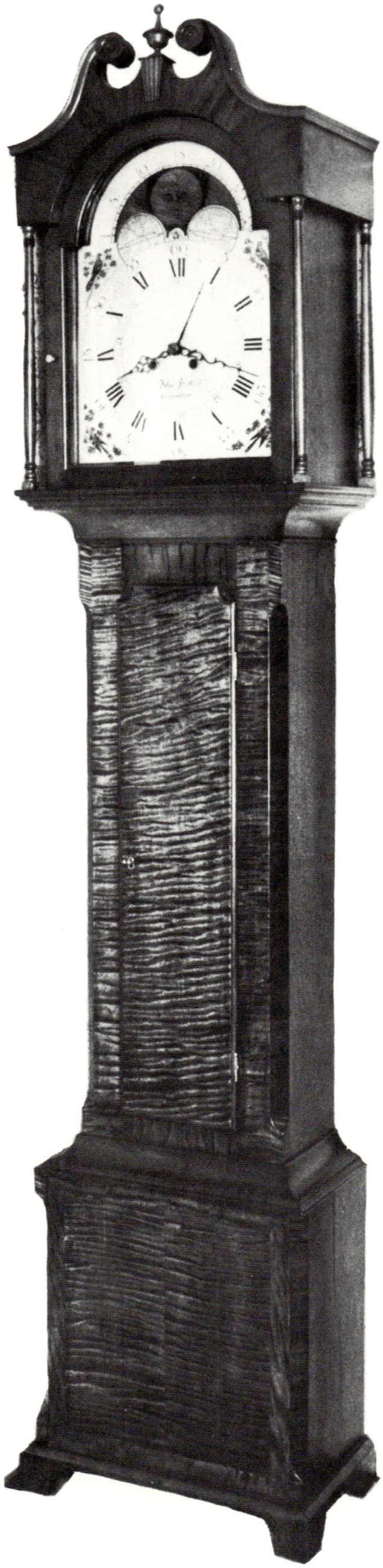

Fig. 2-33. Scroll-pediment case housing eight-day works by John Hoff of Lancaster, dating about 1799 to 1818. The use of curly maple, as seen in the front of the base and waist section of this clock, is unusual in Lancaster County furniture. It is curious that some elements of the hood were not executed in curly maple; the tympanum of the pediment is veneered in mahogany with the grain running on the diagonal from the median. Case made of curly maple, cherry, and mahogany veneers. (Private collection)

Fig. 2-34. Scroll-pediment case housing musical eight-day movement by Martin Shreiner of Lancaster, dating from the beginning of the nineteenth century. This is a good example of early Federal clock cases from Lancaster borough; the matching of figured veneers and crossbanded borders are executed with unusual skill. The lancet or Gothic side windows on the hood are an interesting and uncommon detail. Mahogany primary wood with figured veneers, 101″ high. (Private collection)

Fig. 2-35. Scroll-pediment case housing eight-day works by George Ford of Lancaster, dating about 1808 to 1820. The coloristic treatment of wood, favored in the Federal period, is well illustrated here. The primary wood of the case is cherry, but the front of the base, the pendulum door, and the tympanum of the pediment are veneered in figured mahogany. A surprising note of light color is provided by the curly maple used to fabricate the door to the face. The oval side windows, the roundels which serve as rosettes, and the inverted shell-like motif beneath the center finial should be compared to similar features on the Eberman clock illustrated in Fig. 2-24. 87¼″ high. (Private collection)

Fig. 2-36. Scroll-pediment case housing eight-day works by John Eberman Jr. of Lancaster, probably made just before 1807. The pediment and shaping of the outflaring feet are very similar to comparable features of other clock cases evidently made in Lancaster borough between 1790 and 1820. The delicately reeded corners of base and waist and the inlaid oval paterae on the base and pendulum door are interesting variants. Cherry primary wood with light-wood inlays. (Private collection)

Fig. 2-37. Scroll-pediment case housing thirty-hour works by John Conrad Heinselman, who worked in Manheim from about 1792 to 1803. The treatment of the base with the undulating border defined by light inlay with a central-oval inlaid medallion is very unusual; the form of this inlaid panel may be derived from the raised panels which decorate the base sections of Chippendale-style cases. Note the deep turning of the rosettes. Due to the absence of feet and the possible loss of the original center finial, the original height is not known. Walnut primary wood with light-wood inlays. (Collection Hershey Museum of American Life)

Fig. 2-38. Scroll-pediment case housing thirty-hour movement by Christian Eby of Manheim, probably dating just before 1803. The use of the chevron design stringing to form the base panel with upper hollow corners is echoed in the use of the same inlay to outline the pendulum door. The design is united by using the same inlay in the arch over the door which opens to the face. Cherry with light-wood inlays. (Collection Hershey Museum of American Life)

Fig. 2-39. Scroll-pediment case housing eight-day works by Jacob Eby of Manheim, who worked from about 1803 to 1825. This very well-proportioned Federal case boasts a variety of inlays, the most notable being the eagle at the top of the waist enframed by an oval of patterned inlay. The curve of the scroll pediment complements the cursive movement of the inlaid rosettes. Walnut primary wood with light-wood inlays; the finials may be old replacements, thus altering the original height. (Collection Hershey Museum of American Life)

Fig. 2-40. Scroll-pediment case housing eight-day movement by Jacob Eby of Manheim, probably dating about 1803 to 1825. This richly inlaid bowfront case is one of the great masterpieces of Federal-period cabinetwork from Lancaster County. The eagle inlaid in the oval at the top of the waist is distinguished by its detail and the boldness of its stance. The interlocking ovals of inlay on the base section, enframed by mitered, veneered borders, relate to the figured oval on the pendulum door. The elaborate inlaid shell motif on the center of the skirt is related to inlays favored in Baltimore. Mahogany primary wood with figured veneers and inlays, 105″ high. (Photograph courtesy Henry Francis du Pont Winterthur Museum)

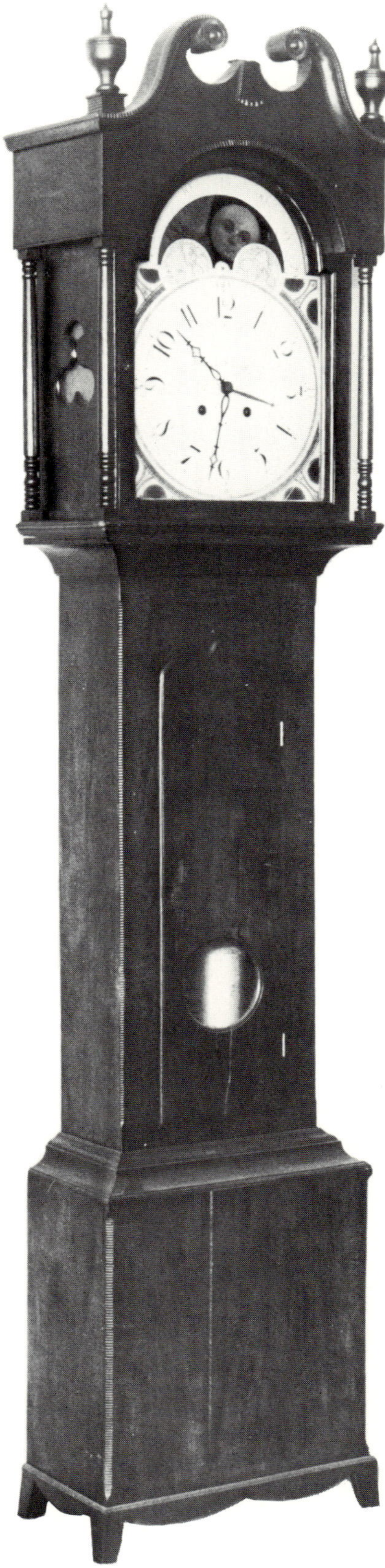

Fig. 2-41. Scroll-pediment case housing eight-day works by Joseph Gorgas of Ephrata, who worked from 1792 to 1805. Certain features of this case, such as the inverted, shell-like motif beneath the center finial, may be related to Federal cases from Lancaster borough. Unusual elements which bespeak an innovative provincial hand are the ribbing in the corners of the base and waist and the serrated edges of the scroll pediment and the arch over the door to the face. The triple-lobed side windows and the glazed round aperture of the pendulum door are also notable. Cherry primary wood, 95½″ high. (Private collection)

Fig. 2-42. Scroll-pediment case housing eight-day movement by John Brown of Lancaster, who worked about 1824 to 1850. Although Brown worked in Lancaster, the great simplicity of this Federal case, plus the very late use of unfluted quarter columns, may suggest the hand of a country joiner. Cherry primary wood, 94¾″ high. (Private collection)

Fig. 2-43. Scroll-pediment case housing eight-day works by George Smith of Strasburg, dating about 1814 to 1827. Although the basic design of this case is typical of many Federal period clocks from Lancaster County, the use of cross-banding for the borders of the waist section's pendulum door and the base, plus the excellent matching of figured veneers in the tympanum of the pediment, is very accomplished. Note the lancet-shaped side windows. Apple primary wood with figured veneers, 96½" high. (Private collection)

Fig. 2-44. Scroll-pediment case housing thirty-hour works by Thomas Dickey of Marietta, dating about 1813 to 1824. This case is distinguished by the attenuation of its general design and details; the high pitch of the scroll pediment accentuates this feeling of verticality. Cherry primary wood, 92" high. (Private collection)

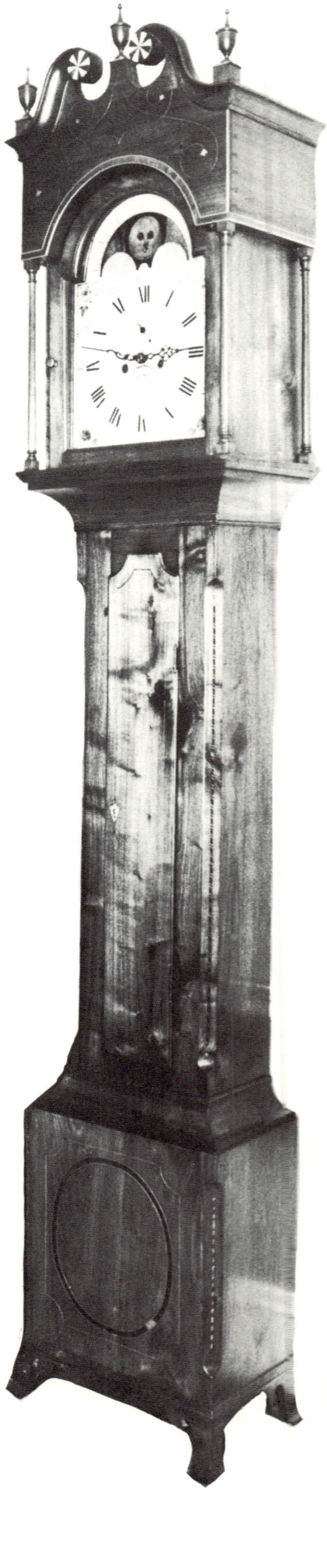

Fig. 2-45. Scroll-pediment case housing eight-day movement by John Esterle of Maytown; on the back of the boards of the waist section this clock is dated 1816. The use of matching figured veneers and broad, mitered borders is very effective. The use of delicate reeding on the chamfered corners, the columns of the hood, the plinths of the side finials, and the keystone beneath the center finial provide a sense of unity. There are no side windows. Cherry primary wood with mahogany veneers, 94¼″ high. (Private collection)

Fig. 2-46. Scroll-pediment case housing eight-day movement by John Davis of New Holland, probably dating just before 1805. The tall, graceful case boasts a wealth of well-executed inlays. The highly stylized vine inlays in the tympanum of the pediment are closely related to the inlays of the case illustrated in Fig. 2-29. The exaggerated shaping of the outflaring bracket feet should be noted. There are no side windows. Walnut primary wood, 104″ high. (Private collection)

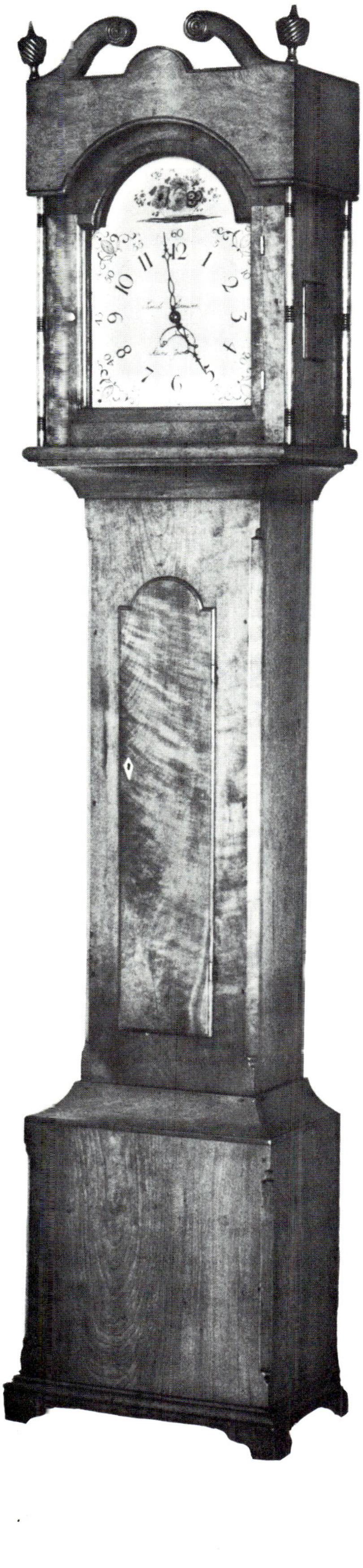

Fig. 2-47. Scroll-pediment case housing thirty-hour movement by Samuel Breneisen of Adamstown, dating about 1800. The very simple treatment of this pediment, without moldings or superfluous details, reduces the typical design of a Lancaster County Federal clock to an almost abstract silhouette. This case tells much of both the ambitions and limitations of a rural joiner. Cherry primary wood, 87¾" high. (Private collection)

Fig. 2-48. Case housing eight-day works by Samuel Quest of Maytown, dating about 1820. The unusual design of this pediment, with six finials, reeded plinths, and curved cresting has been found in several other clock cases from Lancaster County. Note the strips of reeding in the base and the stiles of the waist section. There are no side windows. Cherry primary wood, 95½" high. (Private collection)

Fig. 2-49. Scroll-pediment case housing eight-day movement by Martin, Jr., & Phillip Shreiner of Lancaster dating just after 1835. This illustrates the typical Lancaster County late Federal case, which retains features of earlier Federal cases, but introduces turned feet, broader proportions, and heavier decorative details. On this case, the scalloped-base panel is a remarkable stylistic survival from Chippendale style cases. Cherry primary wood with mahogany veneers, 96″ high. (Private collection)

Fig. 2-50. Scroll-pediment case housing eight-day movement by Anthony Wayne Baldwin of Lampeter Square. Within the waist section this case bears an old note stating that the clock was bought by John Mylin in 1829. There are strong grounds for attributing this case, and others housing works by A. W. Baldwin to Jacob Bachman. Note the small urn finials, graceful curve of the scroll pediment, and lancet side windows. Cherry primary wood with figured veneers, 99½″ high. (Collection Lancaster County Historical Society)

Fig. 2-51. Scroll-pediment case housing eight-day movement by William Fraser of "New Ephrata." The face bears the date October 30, 1839. This is typical of late Federal clock cases from Lancaster County; the use of friezes of dark veneer at both the base and top of the waist section is an unusual detail. The loss of the original feet, and the possible reduction in height of the center finial, have altered the original height. Cherry primary wood. (Private collection)

Fig. 2-52. Scroll-pediment case housing eight-day works by Henry Ober (Jr.) of Elizabethtown, probably dating 1825 to 1835. The use of the half columns at the front of the base, with reeded shafts and a variant of the Ionic Order for the capitals, relates this to Empire furniture styles. The crosshatching on the plinths supporting these half-columns and the three finials is another unusual detail. Cherry primary wood with mahogany veneers, 93¾" high. (Private collection)

Fig. 2-53. Scroll-pediment case housing eight-day movement by Samuel Breneisen of Reamstown; this movement is dated 1834. Although the form of this case is very representative of late Federal cases from Lancaster County, the ornamentation of the British Coat of Arms on the pendulum door and the raised, carved face below the center finial are atypical. The various interpretations of these features are discussed in detail in the text. Cherry primary wood with mahogany veneers, 96½″ high. (Private collection)

Fig. 2-54. Elaborate Victorian case housing eight-day musical movement by Godfried M. Zahm of Lancaster. This clock was made for the Centennial in 1876. This is an unusually late instance of a tall case clock from Lancaster County, and the carved details of the case, particularly the massive central motif of the pediment, bespeak some influence from Eastlake's designs. Walnut primary wood, 100″ high. (Private collection)

Fig. 2-55. Rare mantel or shelf clock with face signed by Samuel S. Grosch of Marietta, probably dating 1820. There is a tradition that the case was made by a member of the Hiestand family in Marietta. The general design of this case was obviously derived from the Terry-type pillar and scroll clocks. However, the shaping of the pediment should be compared to that in Fig. 2-48. Mahogany primary wood with mahogany veneers. (Private collection; photograph courtesy Hovey Gleason)

Fig. 2-56. Wall case housing eight-day movement by William Henry Hall of Lititz, who worked about 1830 to 1868. This unusual clock, with dials visible from both sides of the wall, is believed to have been installed after Hall's death in 1858. Walnut primary wood, 59" high. (Private collection)

Fig. 2-57. Walnut corner cupboard housing thirty-hour movement made by John Hoff of Lancaster. Hoff made this movement for Daniel Sittle of Lampeter Square in 1812. The corner cupboard case may have been made by a joiner in the Lampeter area. Fewer than six corner cupboards with clocks in the central section are known. Walnut primary wood, 92¼″ by 63½″. (Private collection)

Fig. 2-58. Detail of the relief carving on the pediment of the clock illustrated in Fig. 2-16.

Fig. 2-59. Detail of the eagle inlay on the waist of the clock illustrated in Fig. 2-39.

Fig. 2-60. Detail of face motif carved beneath center finial of clock illustrated in Fig. 2-53.

III DIALS

The face of a clock, known as the *dial*, is often taken for granted. As the word "clock," which is derived from the Latin word "clocca" meaning "bell," indicates, the earliest clocks had no dial. Time was indicated by the striking of a bell. As mankind became more time conscious, it was found to be beneficial to affix a dial and *hand*. The hand on European and English clocks made one revolution every twelve hours. Its direction of travel from left to right, from which we get our present-day expression *clockwise*, was probably a result of Northern Hemisphere man previously reckoning time's passage by the path of the sun through the heavens, or of the path of shadows cast on the ground. The hour circle of the clock dial was divided into twelve hour marks. It should be noted that when Roman numerals are used on a dial, IIII is the rule and IV the exception. Each segment between the hours was subdivided into half-, and often quarter-, hours. Due to the poor timekeeping capabilities of clocks before the invention of the pendulum and anchor escapement about 1650, minutes were virtually unknown. Clocks gained and lost time at random and had to be set daily by the sundial. Like many other great moments in history, the date of the application of the dial to the clock is unknown.

A clock dial usually consists of five parts: the *dial plate* which may be either arched or square (other shapes such as round and oval are rare); the *chapter ring* which carries the hour and minute numerals; the *spandrels* which decorate the corners and, on a *solid arched* dial, the arched area; and the *signature* which often includes the maker's town. Other parts frequently found are a *seconds circle*, *date* or *calendar wheel*, the *dial feet*, and the *moon wheel*. The dial feet are riveted to the dial plate and provide a means of attaching the dial to the clock movement. The moon wheel, whose use is generally restricted to eight-day clocks, was an important feature in the eighteenth and nineteenth centuries when streetlights were few and far between, because it allowed for easy computation of the next period of moonlight. Also, certain crops were planted and harvested according to the phases of the moon.

Some parts of the dial differ in construction depending on the type of dial encountered. Between 1750 and 1850 there were three basic types of dial construction: *brass*, *iron*, and *wood*. Brass dials were common to English clocks prior to 1775, whereas iron and wood dials were common to German clocks of the period. No wooden dials have been found on tall clocks made by clockmakers of Lancaster County, and they are not, therefore, discussed here.

THE BRASS DIAL

A brass dial generally consisted of a brass dial plate which had been hardened and polished by hammering (planishing). To the plate were attached the following: brass chapter ring upon which were engraved the hour and minute numerals; cast brass spandrels; seconds circle, and perhaps, the name plate. The chapter ring, seconds circle, and date wheel were often silvered to permit easier reading of the engraved numerals which had been filled with a black mixture consisting of shellac, rosin, and lampblack. Spandrels sometimes were gilt or silvered. Pewter was also used for spandrel castings. The center area of the dial plate within the chapter ring was often *matted* with a hammer and punch, or decorated with engraved designs. The moon wheel might be engraved or painted with the area between the two moon orbs being a blue star-spangled firmament. The *lunar calendar* was silvered and sometimes engraved either on the arch of the dial or on the circumference of the moon wheel. A brass dial properly silvered and polished can be quite a handsome sight.

A modification of the brass dial was the *engraved flat brass dial* that was hand silvered. It is believed to have been introduced in the 1770s, though the exact date is unknown. Thomas Harland (1735–1807) is given credit for being the first to introduce this type dial in his state of Connecticut in 1773. In preparing this book, only two Lancaster County clockmakers utilizing this type of dial were discovered: John Belsner of Ephrata and Wilton Atkinson of Lancaster Borough. The silvered, engraved flat brass dial with its black filler can be extremely pleasing to the eye. It is also much easier to read in a darkened room than a brass dial with attachments.

THE IRON DIAL

The iron dial, common to German clocks of the same period, was similar in design to its contemporary brass dial, however, it was normally square, or if arched, without a moon wheel. The iron dial plate was painted, most often red, black, or white. The attachments (spandrels, chapter ring, etc.) were pewter or German silver. The dial was footless, being screwed onto extended movement plate pillars. The screws also served to hold the chapter ring in place. Early dials by George Hoff are of this type.

In 1772, the Birmingham, England firm of Wilson and Osborne (japanners), brought forth a dial movement that was to change the appearance of the clock dial for much of the following hundred years. This was the *white dial* with its mounting device, the *false plate*. The white dial, with its bright appearance and easy readability, was accepted by clockmaker and buyer alike and very rapidly replaced the brass dial and the German iron dial. The false plate was an important part of the dial package, for once the dial plate had been painted by the manufacturer, it was impossible to relocate the dial feet without damage to the kiln-fired, painted surface. Short dial feet were therefore riveted to the dial plate in the first steps of manufacture and the false plate fitted to them, to be held in place by tapered pins. The clockmaker took a dial of the customer's choosing and attached three or four feet to the false plate in positions where they would not interfere with the operation of the clock movement. He would also drill in the holes for the winding arbors if it was to be an eight-day clock. Grommets were used to cover any damage caused while making the holes.

The false plate has been responsible for much controversy because the name of the dial manufacturer, e.g., Wilson, is often cast into the plate as well as being stamped into the backside of the moon and calendar wheels. This has led many people to erroneously believe that the clock movement, to which the dial and false plate is attached, was also manufactured in England. Dial makers during the first seventy years or so following the invention of the white dial, however, were japanners and not clockmakers.

American clockmakers imported dials from England by the case when trade relations were good. They could order any one of at least three different sizes: twelve, thirteen, or fourteen inches. The dials could be ordered with or without the moon wheel and, eventually, with a choice of Arabic or Roman numerals. The exact date of the introduction of the white dial to the American scene is unknown, but it is known to have been in Philadelphia at the time of the American Revolution and in Baltimore as early as 1784. It was not long before Americans began making their own white dials in order to avoid the import tax on clock parts. Benjamin Witman, a clockmaker in nearby Reading, Pennsylvania, advertised in May 1799 that he was going to "carry on the clock dial manufactory in all its various branches" in addition to clock and watchmaking. Lancaster's Philip Schaeffer and Peter Lehn Grosh (a portrait painter) advertised white dials as did the Harrisburg merchants Ogelsby & Pool. Many early white dials and false plates that do not carry the name of a manufacturer also often do not have the polished look of the English dial and, together with their comparatively thin sheet-iron false plates, are probably American made. It is unfortunate that so few American dialmakers put their names on their products.

Once the clockmaker had fitted the dial to his clock movement, he could have the dial marked with his name and the name of his town. He might do the lettering himself or have a local artist do it for him. The lack of a maker's name was often due to the wishes of the buyer. At times, however, it was due to the lack of someone to do the lettering. It should be remembered that many people were illiterate at this time and some clockmakers could be numbered among them. A variation in the spelling of a clockmaker's name (e.g., *Esterly* instead of *Esterle*) can easily be explained as being the result of an artist using his own judgment in the spelling. Some signatures, such as that of Lancaster Borough's Martin Shreiner, were painstakingly done and are, in themselves, beautiful. It is interesting to note that the day book of Lancaster's famous portrait painter, Jacob Eichholtz, carries two entries of lettering done for George Hoff, leading to speculation that Eichholtz may have done some of Hoff's dial lettering. Often, German lettering was used and this tends to confuse those who are not familiar with it, causing them to misread the maker's name.

The spandrel design of the white dial went through an evolution from a simple simulation of the old cast spandrels to more decorative and colorful flowers, birds, seashells, and geometric designs. In England, and on a lesser scale in America, representation of the seasons or other vignettes also appeared. It should be pointed out that the spandrels (and moon wheels) were hand painted so that although one might find similar dials, one would not expect to find identical ones.

Clock hands also went through an evolution, progressing from unmatched to matched iron hands (about 1780) and then to matched brass hands (about 1805). Pewter hands were also used. Various styles of hands are represented here, some quite exotic.

The following dials were chosen to represent not only the different types discussed above, but also to show varying spandrel designs, hands, and signatures. The dials are described by width, construction, and special features. The dialmaker's name is indicated when the false plate, moon wheel, or calendar wheel bears his name. Brass dials are believed to have been made by the clockmaker himself. Dials by the following Birmingham, England dialmakers have been found during this project: Thomas Keeling, who worked from 1800 to 1815; Thomas Hadley Osborne, who worked initially with James Wilson from 1772 to 1777, then alone until his death about 1779, to be succeeded by his wife Anne through 1798; Walker & Hughes, at work from 1815 to 1835; and James Wilson, as previously mentioned, with T. H. Osborne from 1772 to 1777, then alone until his death in 1809.[1]

1. Brian Loomes, *The White Dial Clock* (New York: Drake, 1975).

3-1. Wilton Atkinson 12″ arched, engraved flat brass dial ca. 1773. Silvered. Painted moon and star-spangled firmament. Lunar calendar engraved on the moon wheel. Footed dial plate. (Private collection)

3-2. Hamilton Ayars 14″ arched white dial ca. 1840. False plate. (Private collection)

3-3. Anthony Wayne Baldwin 14″ arched white dial ca. 1829. Sweep seconds and calendar hands. False plate. (Lancaster County Historical Society)

3-4. John Belsner 11" square engraved flat brass dial ca. 1780. Silvered. Dial screwed to four extended movement pillars. (Private collection)

3-5. Joseph Bowman (Jr.) 14″ arched white dial ca. 1840. False plate. (Mill Bridge Craft Village/Lefever Historical Society)

3-6. Samuel Breneisen 12″ solid arched white dial ca. 1800. No false plate. (Private collection)

3-7. Samuel Breneisen 14″ arched white dial ca. 1834. False pla e. (Private collection)

3-8. John Brown 12″ solid arched white dial ca. 1830. False plate. (Private collection)

3-9. Thomas Bradin Burrowes 13″ arched white dial ca. 1800. Sweep seconds hand. No false plate. (Private collection)

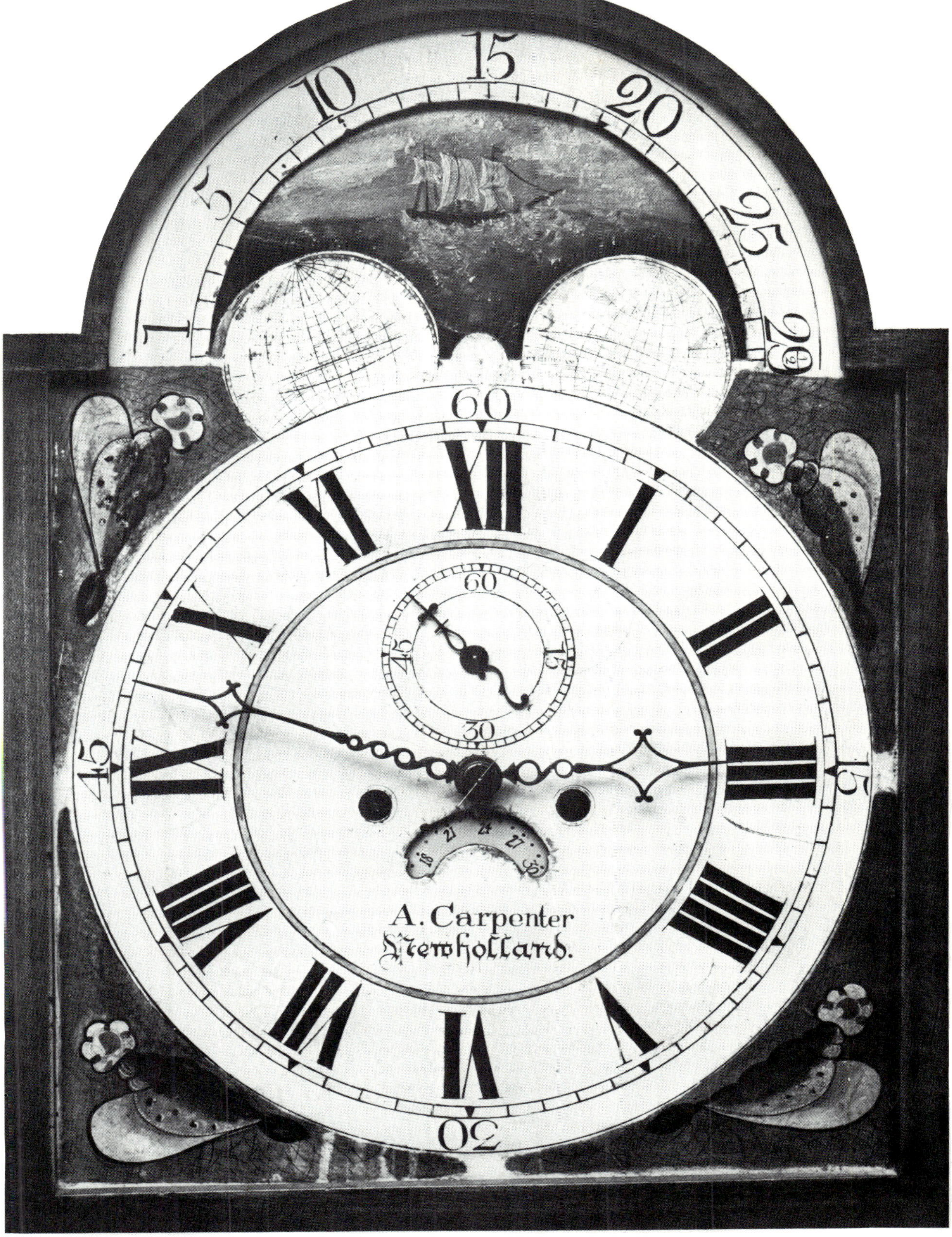

3-10. Anthony Carpenter 14″ arched white dial ca. 1830. False plate.
(Private collection)

3-11. Anthony Wayne Carpenter 14″ arched white dial ca. 1835. False plate. (Private collection)

3-12. John Davis 13″ solid arched white dial ca. 1800. False plate marked "Osborne." (Private collection)

3-13. Thomas Dickey 12″ solid arched white dial ca. 1820. No false plate. (Private collection)

3-14. George David Eberman 14″ arched white dial ca. 1840. False plate. (Private collection)

3-15. John Eberman, Jr. 12″ arched brass dial ca. 1785. Matted dial plate and engraved arch. Painted moon and star-spangled firmament. Lunar calendar engraved on moon wheel. Spandrels missing. (Private collection)

3-16. John Eberman, Jr. 12″ solid arched white dial ca. 1790. No false plate. (Private collection)

3-17. John Gottlieb Eberman 13″ arched white dial ca. 1790. Sweep calendar hand. False plate marked "Wilson." (Private collection)

3-18. John Joseph Eberman 13″ solid arched white dial ca. 1810. No false plate. (Private collection)

3-19. Christian Eby 12″ solid arched white dial ca. 1800. No false plate. (Hershey Museum of American Life)

3-20. Jacob Eby 13″ arched white dial ca. 1805. Sweep seconds and calendar hands. False plate marked “Osborne.” (Hershey Museum of American Life)

3-21. John Erb 14" arched white dial ca. 1835. Sweep seconds (missing) and calendar hands. False plate. (Private collection)

-22. John Esterle 14″ arched white dial ca. 1800. predating the clockmaker. Sweep seconds and calendar hands. False plate. (Private collection)

3-23. John Esterle 14″ arched white dial ca. 1830. False plate. (Private collection)

3-24. George Fisher 13″ arched white dial ca. 1800. Sweep seconds and calendar hands. No false plate. (The Amish Farm and House)

3-25. George Fisher 12″ solid arched white dial ca. 1800. No false plate. (Private collection)

3-26. George Ford 13″ arched white dial ca. 1810. False plate. (Private collection)

3-27. Christian Forrer 11″ arched brass dial ca. 1770. Painted moon and star-spangled firmament. Matted dial-plate center and engraved arch. Silvered spandrels. Sweep seconds hand (pointer missing). (Virginia Museum of Fine Arts)

3-28. Christian Forrer 12″ solid arched brass dial ca. 1770. Brass spandrels. Engraved medallion. (Private collection)

3-29. William Fraser 14″ arched white dial ca. 1839. False plate. (Private collection)

3-30. Jacob Gorgas 11″ square iron dial ca. 1770. Pewter chapter ring and Georgian style spandrels. Alarm setting disc. (Private collection)

3-31. Jacob Gorgas (Jr.) 12½″ solid arched brass dial ca. 1790. Engraved dial center. Silvered chapter ring, medallion, and calendar wheel. (Private collection)

3-32. Jacob Gorgas (Jr.) 12″ solid arched white dial ca. 1795. Calendar wheel stamped "Wilson–Birmm." (Private collection)

3-33. Joseph Gorgas 14″ arched white dial ca. 1800. Sweep seconds (missing) and calendar hands. False plate. (Private collection)

3-34. Solomon Gorgas 12″ solid arched white dial ca. 1800. Dial is screwed to the extended movement pillars. Dial very similar to another Solomon Gorgas dial not included here. Both possibly locally manufactured. (Private collection)

3-35. Christian Hall 13" solid arched white dial ca. 1800. "T. Keeling" marked on the false plate. (Private collection)

3-36. Christian Hall 14″ arched white dial ca. 1825. Sweep seconds and calendar hands. False plate. (Private collection)

3-37. William Henry Hall 10″ painted glass dial ca. 1865 Sweep seconds hand. Probably painted by Hall himself. (Private collection)

3-38. John Conrad Heinselman 13″ arched white dial ca. 1790. Sweep seconds and calendar hands. No false plate. (Private collection)

3-39. John Conrad Heinselman 13″ arched white dial ca. 1800. Sweep calendar hand. No false plate. (Hershey Museum of American Life)

3-40. John Hoff 12″ solid arched white dial ca. 1800. No false plate.
(Private collection)

3-41. John Hoff 14″ arched white dial ca. 1800. Sweep calendar hand. False plate. (Private collection)

3-42. John Hoff 14″ arched white dial ca. 1800. Sweep seconds and calendar hands. This dial is mounted on a quarter-hour chime movement made in 1810. False plate marked "Osborne." (Lancaster County Historical Society)

3-43. John George Hoff 11" square iron dial ca. 1770. Dial plate painted red. Attached pewter spandrels, chapter ring, and name plate. Dial is screwed to extended movement pillars. (Collection of the Pennsylvania Farm Museum of Landis Valley)

3-44. John George Hoff 10″ solid arched, footed, iron dial ca. 1770. Pewter spandrels and chapter ring. Note asymmetrical winding holes and crossings of the Roman numerals. (Lancaster County Historical Society)

3-45. John George Hoff 11″ arched brass dial ca. 1785. Footed. Painted moon and star-spangled firmament. Lunar calendar engraved on the moon wheel. Engraved arch and dial center. Pewter spandrels and German silver chapter ring. (Private collection)

3-46. John George Hoff 12″ solid arched brass dial ca. 1790. Footed. Pewter spandrels. Sweep seconds hand. (Private collection)

3-47. John George Hoff unsigned 12″ solid arched white dial ca. 1800. Rocking ship. Perforated strap dial feet are pinned to extended movement pillars. Note off center winding holes common to Hoff's eight-day movements. Dial of possible Lancaster County manufacture. (Hershey Museum of American Life)

3-48. John George Hoff 14″ arched white dial ca. 1800. Sweep seconds hand. False plate and moon wheel marked "Wilson–Birmm." (Private collection)

3-49. John George Hoff 14" arched white dial ca. 1810. Sweep calendar hand. This dial mounted on a ten-tune musical/quarter-hour strike movement. Musical selections are manually selected by a hand mounted at the moon wheel pivot. This dial has been restored. (Private collection)

3-50. John George Hoff, Jr. 13″ solid arched white dial ca. 1810. No false plate. (Private collection)

3-51. John George Hoff, Jr. 14″ arched white dial ca. 1815. False plate and moon wheel marked "Walker & Hughes–Birmingham." (Private collection)

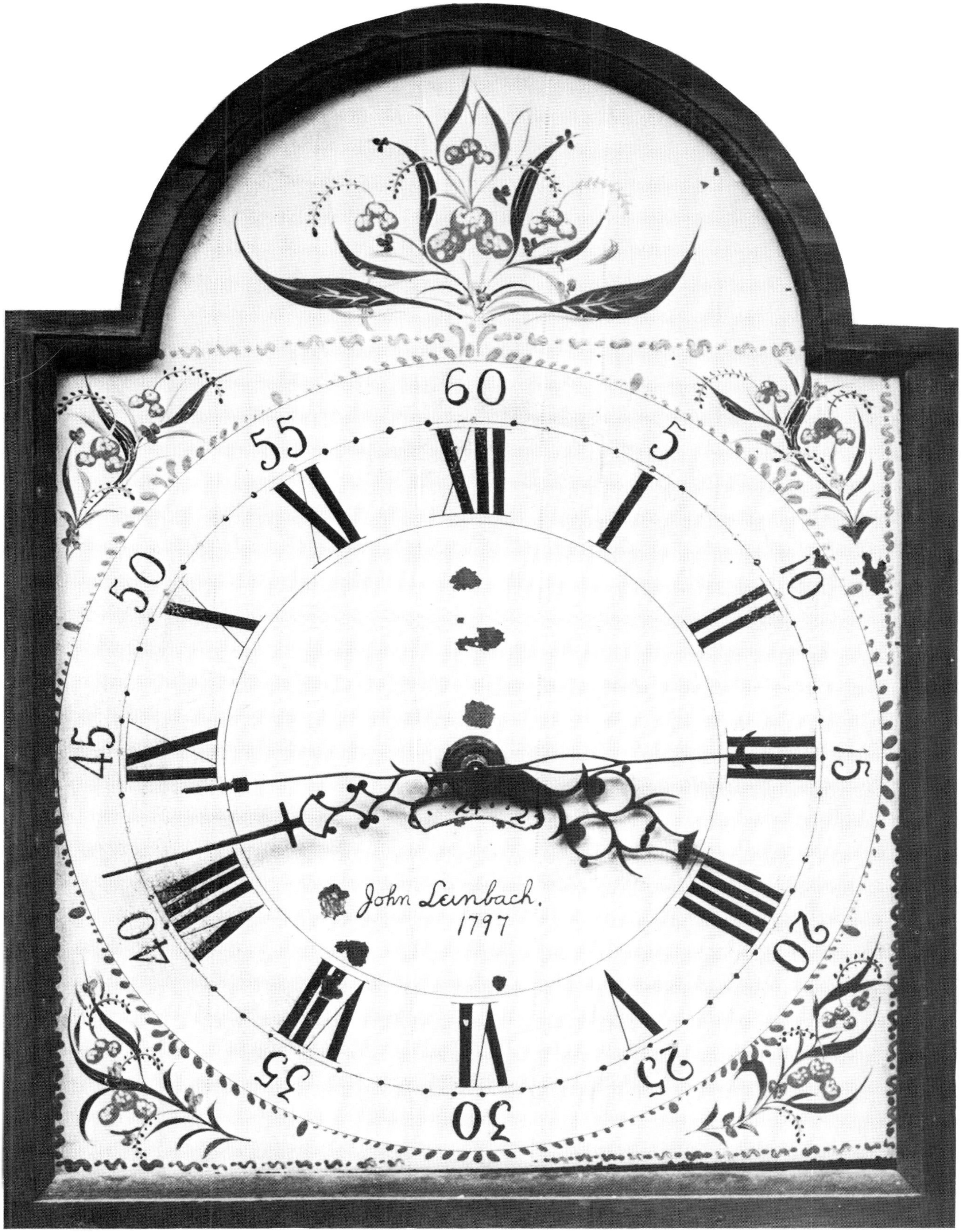

3-52. John Leinbach 12″ solid arched white dial ca. 1795. A thirty-hour clock with a rare sweep seconds hand. No false plate. (Private collection)

3-53. Henry Lewis Montandon 14" arched white dial ca. 1800. "Osborne's Manufactory Birmingham" marked in false plate and stamped in moon wheel. (Collection of Richard W. Case)

3-54. Henry Ober (Jr.) 14″ arched white dial ca. 1830. False plate. (Private collection)

3-55. Samuel Quest 13″ arched white dial ca. 1820. False plate. (Private collection)

3-56. Samuel Quest 14″ arched white dial ca. 1820. False plate. (Private collection)

3-57. Martin Shreiner 12″ solid arched white dial ca. 1795. No false plate. Similar spandrel design on Shreiner's clocks Nos. 75, 85, and 90 suggest a common dialmaker. (Collection of the Pennsylvania Farm Museum of Landis Valley)

3-58. Martin Shreiner 14" arched white dial ca. 1810. This dial mounted on a seven-tune musical movement. False plate. Sweep seconds and calendar hands. (Private collection)

3-59. Martin (Jr.) & Philip Shreiner 14″ arched white dial ca. 1835. Sweep seconds and calendar hands. False plate. (Private collection)

3-60. George Smith 14″ arched white dial ca. 1800. False plate marked "Osborne." Dial predates the clockmaker. (Private collection)

3-61. Samuel C. Stauffer 11½″ arched brass dial ca. 1795. Painted moon and star-spangled firmament. Engraved lunar calendar and dial center. Pewter chapter ring and brass spandrels. (Hershey Museum of American Life)

3-62. Rudy Stoner 10½″ arched brass dial ca. 1760. Pewter spandrels. Matted dial center. Dial screwed to extended movement pillars. (Private collection)

3-63. Abel Witwer 12″ solid arched white dial ca. 1795. No false plate. (Private collection)

3-64. Godfried M. Zahm 13½″ brass dial. Silvered. Sweep seconds hand. Separate days of week, month, and calendar circles. This dial is mounted on Zahm's Centennial musical clock. (Private collection)

3-65. Godfried M. Zahm Centennial musical clock moon dial. Dial is located in the trunk door. (Private collection)

IV MOVEMENTS

The works, more properly called *movements*, of the Lancaster County clockmakers are of two basic designs: that which follows in the more common English tradition, and that which follows in the German tradition. To properly appreciate the differences between these two traditions, it is first necessary to understand just what makes a clock tick.

MOVEMENTS IN THE ENGLISH TRADITION

A clock movement is a mechanical device designed to release energy at a controlled rate. Time is indicated either by releasing periodically a separate similar energy source causing a bell to be struck or by the position of a hand or hands on a *dial*, or by both means. The eight-day clock movement normally encountered in tall clocks (now commonly referred to as "grandfather clocks") made between 1750 and 1850 consists of two sets of *trains* (gears) mounted between two vertical brass plates. The plates are separated by four or more brass *pillars* that are normally riveted to the *back plate* and extend far enough through holes in the *front plate* to permit them to be secured by tapered pins. Within this framework are the *time train* and the *strike train*. Each train consists of large brass *wheels* and smaller steel wheels properly called *pinions*. Both wheels and pinions are mounted on steel *arbors* whose reduced ends, the *pivots*, turn in *pivot holes* drilled in the front and back plates. The wheels are driving, the pinions are driven. The closest wheel to the power source (normally an eleven- to thirteen-pound iron or, now fairly rare, lead *weight*) is the largest wheel and has the greatest number of teeth in the train. The movement is normally designed so that this wheel, known as the *great wheel*, makes one revolution every twelve hours. Adjacent to this wheel is a *ratchet wheel* and *winding drum*. The *gut*, cable, or cord is wound up on the drum. The drum may be either smooth or grooved, the latter helps to ensure proper lapping of the gut. The great wheel, ratchet wheel, and drum are all mounted on the *winding arbor*. Winding is done weekly with a *crank*.

Usually, the farther away a wheel is from the power source, the smaller is its diameter and the number of its teeth, and, therefore, the faster it revolves. The pinions normally have only six to eight teeth that are known as *leaves*. The *center arbor* of the time train is designed to make one revolution per hour and it is its extended pivot that drives the *minute hand*. The fourth arbor carries the *escape wheel* that in most clocks of this period releases, or allows to escape, one unit of energy every second. It is on its extended pivot that one often finds a small *seconds hand*.

The device that allows the energy to escape is the *anchor*, so called because its surfaces, the *pallets*, resemble the flukes of a ship's anchor. There are basically two types of escapement used: the common *anchor/recoil* where a slight backwards movement of the seconds hand is noticeable, and the *dead-beat* where the seconds hand is seen to advance in even on/off steps. The anchor is mounted on the arbor farthest from the power source, the *pallet arbor*. The back of the pallet arbor passes through a large hole in the back plate and is supported by the two-footed *bridge* that also acts as a suspension point for the *pendulum*. The upper end of the pendulum is a piece of spring steel, the most fragile part of the movement. It is the pendulum with its circular, brass-sheathed *bob* that regulates the frequency of the escape of energy and thus determines, under ideal conditions, the accuracy of the clock. The distance between the point of suspension and the center of gravity of the bob is normally approximately thirty-nine inches or a meter long. Increasing the length by turning the *rating nut* to the left will slow the release, whereas decreasing the length by turning the rating nut to the right will increase the release. The pendulum is joined indirectly to the pallet arbor by the *crutch wire* with its forked end.

Outside the movement, and mounted on the front plate, are the wheels and pinions that make up the *motion work* and carry the minute and hour hands. Although no examples are known to have been made in Lancaster County, early English and Philadelphia clocks often have only one hand, the hour hand. The motion work differs from the time train in that the pinions are normally brass like the wheels and are driving, whereas the wheels are driven. Thus the faster turning minute hand pipe is driving, through an intermediate wheel and pinion, the slower hour hand pipe. This train and its motion work when not associated with a strike train is technically known as a *timepiece* and not a clock, the term clock coming from the Latin word for bell. The movement attributed to William Henry Hall (Fig. 4-52) is an example of a timepiece.

When part of a clock, the time train is normally

4-1. Joseph Gorgas Sweep seconds eight-day brass tall clock movement with front plate removed to show strike train (left) and time train (right). (Private collection)

found to the right as one faces the dial. The strike train, to the left, also consists of a series of wheels and pinions but its final arbor carries a two-vaned *fly* that serves as a governor and prevents the blows on the bell from being so fast as to be nearly indistinguishable. The blows are delivered by a steel hammer that is activated by approximately eight steel pins driven into the circumference of the second wheel of the train. The number of blows is controlled by an automatically correcting indexing device known as a *rack and snail.* This device is located outside the movement on the front plate. The snail, a series of twelve steps, each of a diminished radius, is normally attached to the *hour pipe* which also carries the hour hand. A pin to activate the auxiliary *calendar wheel* or *calendar hand* and *moon wheel*, may be imbedded in the snail. The strike

4-2. Time train Fig. 1. (Private collection)

4-3. Strike train of Fig. 1. (Private collection)

operation involves *warning* in which the rack is allowed to fall until its appendage, the *rack tail*, comes in contact with the snail. Simultaneously the fourth wheel, the *warning wheel*, makes a half revolution until it is stopped by contact between a single steel pin driven into its circumference and the *warning lever*. The train remains cocked until the *lifting detent* falls off the *hour wheel lifting pin*. The train, once freed, runs until the single tooth, known as the *gathering pallet*, on the extended pivot of the third arbor, has "gathered up" the number of teeth of the rack made available by the snail. Each revolution of the gathering pallet is accompanied by a blow of the hammer on the bell. The train is locked when the tail of the gathering pallet comes to rest on a pin at the end of the gathered up rack.

Clocks with a long, *sweep*, seconds hand (uncommon in England) have a slightly different time train layout so that the escape arbor can be centrally located. Motion work is driven from the side instead of the center. A sweep calendar hand from the center (also uncommon in England) is normally driven by an intermediate wheel.

Chiming and musical clocks will usually have one or more additional trains. Thirty-hour tall-clock movements have one less arbor in each train and utilize a pulley, through which runs a rope or chain, instead of a drum. When an *endless chain* or *endless rope* is used, only the pulley of the strike train is ratcheted. This allows the single weight to continue to provide power to the time train during winding, thus providing a form of *maintaining power*. With a *two-weight pull-up*, each pulley has its own ratchet. Thirty-hour weights range from approximately four pounds apiece for a two-weight pull-up to seven pounds for the single weight of an endless chain. The reason is that the fewer the wheels, the less friction there is to be overcome and hence the less weight required. Also, in order to keep the required fall of the eight-day weight within reason, the weight is normally attached to a pulley and the end of the gut is attached to the *seatboard* upon which the movement sits. The weight is said to be compounded and requires more mass than a weight directly attached to the end of the gut.

The strike indexing of an English thirty-hour movement differs from that of the eight-day in that it is not directly connected to the hour pipe and, therefore, is not self-correcting. Instead, a *count wheel* with cuts in it at progressively greater intervals allows the train to strike until a *locking* or *count lever* falls into the next cut and causes an appended locking lever to stop the train.

The foregoing has been a description of eight-day and thirty-hour tall clock movements normally encountered. What might be normal for clocks made in Philadelphia, New Jersey, and elsewhere in the colonies where clockmakers trained in the English tradition handcrafted their clocks and trained their apprentices in a similar manner, was not normal or usual in the frontier counties of Pennsylvania where immigrant German clockmakers took up residence.

MOVEMENTS IN THE GERMAN TRADITION

In its extreme, the German tradition differs from the English tradition in a number of ways. In earlier movements iron is used for the pillars instead of brass. Iron was also used for the single-footed pendulum *cock*. Instead of the top of the pendulum rod ending in a piece of spring steel, there was a hook and the pendulum hung on a thread loop. *Wire pinions* might also be utilized. Wire pinions take one of two forms. *Trundles*, pieces of wire, are driven into small holes that have been drilled in the *collet* upon which the wheel is mounted. If the pinion is capped with a *shroud*, it is called a *lantern pinion* after the antique form of lantern that it resembles. If it is without the shroud, it is known as an *open-ended pinion*. Lancaster Borough's master clockmaking Hoff family utilized these pinions in both forms. Wire pinions allow dirt to fall free of a dry pinion whereas pinions made from *pinion wire* which has been drawn through a series of dies, or cut by hand, permit dirt to become impacted in its leaves. The former are also easier to repair than the latter. Wire pinions came into common use in America with the Yankee brass shelf clock movement.

The manner of locking the strike train also differs from English design of the period in that a tailless gathering pallet is used. Locking takes place between the plates where an appendage to the *rack hook* engages the warning wheel pin after the final blow of the strike hammer. Thirty-hour brass-movement tall clocks, although fairly rare outside of Pennsylvania, are quite prevalent within the Commonwealth. Whereas English thirty-hour movements use the *crossings* or spokes of the wheels as the ratchet, Pennsylvania movements have a ratchet wheel mounted on the pulley similar to that on the drum of an eight-day clock. Another characteristic of the Pennsylvania thirty-hour movement is the use of rack and snail striking instead of the count wheel. Levers of movements in the German tradition tend to be heavier and more angular than those in the English tradition. It should be noted that there is evidence that clockmakers who immigrated from Switzerland (e.g., Christian Forrer) worked in the English rather than German tradition.

Various modifications and combinations of the two traditions were employed by the Lancaster County clockmakers. Other than the Hoff family, only Frederick Heisely and Joseph Gorgas are believed to have used wire pinions, and then not consistently. Martin Shreiner (Sr.) advertised that he would, on request, make movements using them, although no examples have been located. Many of the County's clockmakers used the German strike train locking system with its tailless gathering pallet, a shape seldom found on English clock movements of the same period.

4-4. Front plate of Fig. 1 showing the motion work and strike levers. (Private collection)

In 1750 clockmakers were required by necessity to make all the parts by hand. By 1795 complete English movements could be obtained from importers. The import duties imposed on clocks, watches, and parts for both entering the port of Philadelphia ranged from 17.5 percent in 1804 to 37.4 percent in 1812. Philip Schaeffer, a Lancaster Borough merchant, advertised English brass movements for sale in 1796. Clock glass and tools were among the items sold at public auction following the death of yet another Lancaster Borough merchant, Charles Heinitsch, in 1803. An entry in the Heinitsch ledger shows that clockmaker John Eberman bought clock pinions from him. At the auction previously mentioned, Mathias Zahm, the father of G. M. Zahm, purchased eleven "Cuckus" for five-and-one-half shillings. Jacob Gorgas (Jr.) bought chain wire and pinions from the firm of Ogelsby & Pool in Harrisburg. Henry Ober (Jr.) bought weights and brass from the same firm. Wheel blanks, bells, gut, and other items were also available from local merchants, but it was still up to the clockmaker to turn these items into a clock movement reflecting his own design characteristics. Surviving inventories of the estates of clockmakers of the County include the necessary tools and engines of the trade such as wheel-cutting engines, drum-grooving tools, and even a rare fusee engine used in making spring-powered clocks.

Regardless of the degree of manufacture required on the clockmaker's part, he was still a clockmaker and seldom also a cabinetmaker. Often he was also a jeweler, silversmith, engraver, watchmaker, bucklemaker, sealmaker, or maker of surveyors' or mathematical instruments. A number of the County clockmakers also kept tavern.

It should be pointed out that the appearance of a name on a clock dial does not necessarily imply that the person made any or all of the clock movement behind the dial. On occasion the buyer might have his own name engraved or painted on the dial. A cabinetmaker might also place his name on an unsigned dial. Conversely, an unsigned dial does not imply that the movement behind it is an import: the clock order book of Lancaster Borough's John Hoff contains entries for clocks with dials to be "marked" which would suggest that those orders not calling for a marked dial might be left unsigned.

The majority of the clock movements made in Lancaster County during this period were eight-day bell strike brass movements for tall clocks. A lesser number of thirty-hour movements were made. On occasion, one finds a quarter-hour striking, quarter-hour chiming, or musical tall clock. Few chamber (table) clocks were made, and even fewer wall clocks. Although some Connecticut shelf clocks (e.g., *Seth Thomas*, *Board-*

4-5. Close-up view of a lantern pinion of a George Hoff, Jr. thirty-hour movement. (Private collection)

4-6. Eighteenth-century tinned lantern from which the lantern pinion gets its name. (Pennsylvania Historical and Museum Commission, Ephrata Cloister)

man and Wells, etc.) with wooden or stamped brass movements have been discovered with paste-over labels of local clockmakers, no known County-made wooden or stamped-brass movements have been found. Only one signed movement using iron plates with brass bushings instead of brass plates has been located.

An eight-day brass tall clock movement with a thirteen-inch dial with moon wheel sold for £17 5s. ($46.00) in 1807. A thirty-hour clock sold for £9 less. A movement with gilt hands and a sweep seconds hand cost an additional 15s. Quarter-hour four-bell chiming clocks with their additional train were approximately £10 more than the standard eight-day bell strike. In 1810, John Hoff constructed a ten-tune musical eight-day clock that also struck quarter-hours for £80. One can easily see that clocks were fairly expensive items. The case, which was normally purchased by the buyer from the cabinetmaker of his choice, added another £10 or so to the overall price.

4-7. Jacob Gorgas's wheel cutting engine. The wheel blank is bolted down on the vertical spindle. A cutter on another spindle is placed between the yoke of the cutter frame and hand driven. The proper number of teeth to be cut will be found listed on one of the many concentric circles of the index plate. The index pin of the index arm is placed in one of the holes of the selected circle. After each cut of the wheel is made, the pin is lifted clear and the index plate turned to the next hole. (Earl T. Strickler Collection)

The photographs that follow are arranged in alphabetical order according to the name carried on the clock dial to enable easier location of an example by a given clockmaker. When more than one movement by a given maker is shown, the order is thirty-hour, eight-day, quarter-hour strike or chime, and musical. The view generally presented is of the left rear quarter (right for thirty-hour clocks) as there are more visible features in the strike train. Although the majority of the photographs were made with the movement on a stand, a few, of necessity, were taken with the movement still in the case. As photography was done "on location," dials were generally not removed. What one sees, then, is what would be seen if the bonnet of a clock were removed.

It is hoped that a comparison of movements attributed to known Lancaster County clockmakers herein with other movements attributed to the same maker will provide further evidence of a given clockmaker's design characteristics. Perhaps a careful comparison with a movement behind an unsigned dial will establish a possible identity of that movement's maker.

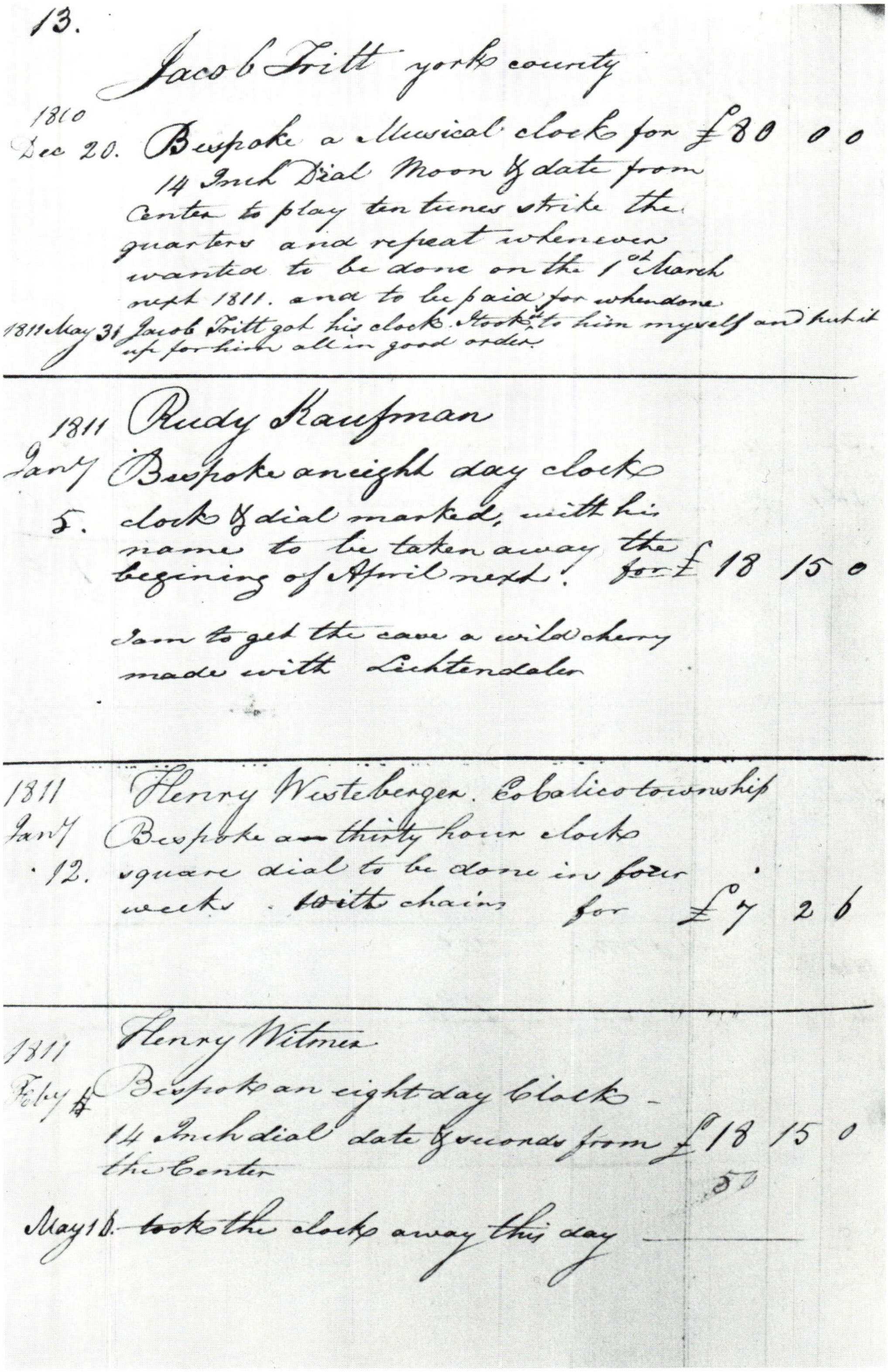

13.

Jacob Fritt york county

1810 Dec 20. Bespoke a Musical clock for £80 0 0
14 Inch Dial Moon & date from
Center to play ten tunes strike the
quarters and repeat whenever
wanted to be done on the 1st March
next 1811. and to be paid for when done
1811 May 31 Jacob Fritt got his clock. I took it to him myself and put it
up for him all in good order.

1811 Rudy Kaufman
Jany 5. Bespoke an eight day clock
clock & dial marked, with his
name to be taken away the
begining of April next. for £18 15 0

I am to get the case a wild cherry
made with Lichtendaler

1811 Henry Westeberger. Cocalico township
Jany 12. Bespoke a thirty hour clock
square dial to be done in four
weeks. with chains for £7 2 6

1811 Henry Witmer
Feby 4 Bespoke an eight day Clock -
14 Inch dial date & seconds from £18 15 0
the Center
50
May 18. took the clock away this day

4-8. Page 13 from John Hoff's "Book of new Clocks made and Sold–Lancaster 1800–" (Lancaster County Historical Society)

4-9. Wilton Atkinson eight-day brass movement ca. 1773. Anchor recoil escapement. Smooth drums and tailless ratchet wheel clicks. Open-ended crutch fork. Repeat-on-demand rack-and-snail striking with gathering pallet and rack-pin locking. "Ears" on the front plate carry the brass dial. Another extant eight-day Atkinson movement has a sweep seconds hand. (Private collection)

4-10. Hamilton Ayars eight-day brass movement ca. 1840. Anchor recoil escapement. Smooth drums with tailed ratchet wheel clicks. Closed crutch fork. Nonrepeat rack-and-snail striking with gathering pallet and rack-pin (large, dial side) locking. Fly has vanes with concave ends. (Private collection)

4-11. Anthony Wayne Baldwin eight-day brass movement made in 1829. Sweep seconds and calendar. Anchor recoil escapement. Smooth drums and tailed ratchet wheel clicks. Closed crutch fork. Repeat-on-demand rack-and-snail striking with gathering pallet and rack-pin (dial side) locking. Graceful, fully polished levers. Bell marked "John Wilbank Caster Philada." This clock was originally purchased for $46.00 in 1829. (Lancaster County Historical Society)

4-12. John Belsner thirty-hour movement ca. 1780. Anchor recoil escapement. Iron plates bushed with brass. Endless chain. Strike train pulley with ratchet wheel and tailless click. Iron pendulum suspension cock with thread loop. Closed crutch fork. Repeat-on-demand rack-and-snail striking with rack-hook locking. Long pivots. Dial carried on extended plate pillars. (Private collection)

4-13. Joseph Bowman (Jr.) eight-day brass movement ca. 1840. Anchor recoil escapement. Closed crutch fork. Smooth drums. Non-repeat rack-and-snail striking with gathering pallet and rack-pin (movement side) locking. Rectangular segment removed from bottom edge of both plates. (Mill Bridge Craft Village/Lefever Historical Society)

4-14. George Breneisen thirty-hour brass movement ca. 1825. Anchor recoil escapement. Endless chain. Open-ended crutch fork. Repeat-on-demand rack-and-snail striking with rack-hook locking. (Private collection)

4-15. Samuel Breneisen thirty-hour brass movement ca. 1810. Anchor recoil escapement. Endless chain. Open-ended crutch fork. Repeat-on-demand rack-and-snail striking with rack-hook locking. (Private collection)

4-16. Samuel Breneisen eight-day brass movement ca. 1834. Anchor recoil escapement. Open-ended crutch fork. Smooth drums with tailed ratchet wheel clicks. Repeat-on-demand rack-and-snail striking with rack-hook locking. Only three plate pillars. Front plate marked "S. E. 1834 1" leading to a theory that the movement was actually made by Samuel, Jr. who was by trade a tailor. (Private collection)

4-17. John Brown eight-day brass movement ca. 1825. Anchor recoil escapement. Closed crutch fork. Grooved drums and tailed ratchet wheel clicks. Extended escape arbor providing for short seconds hand although the dial is not drilled for this feature. Rack-and-snail striking with gathering pallet locking. Narrow bell hammer. Movement attached to the seat board at the bottom edge center of both plates. (Private collection)

4-18. Thomas Bradin Burrowes eight-day brass movement ca. 1800. Sweep seconds. Dead-beat escapement. Open-ended crutch fork. Smooth drums and tailless ratchet wheel clicks. Rack-and-snail striking with gathering pallet and rack-pin (dial side) locking. Tapped bell hammer. (Private collection)

4-19. Anthony Carpenter eight-day brass movement ca. 1830. Anchor recoil escapement. New crutch fork. Grooved drums with tailless ratchet wheel clicks. Repeat-on-demand rack-and-snail striking with gathering pallet and rack-pin (dial side) locking. (Private collection)

4-20. Anthony Wayne Carpenter eight-day brass movement ca. 1835. Anchor recoil escapement. Closed crutch fork. Grooved drums and tailed ratchet wheel clicks. Nonrepeat rack-and-snail striking with gathering pallet and rack-pin locking. (Private collection)

4-21. John Davis eight-day brass movement ca. 1800. Anchor recoil escapement. Open-ended crutch fork. Grooved drums with tailed ratchet wheel clicks. Repeat-on-demand rack-and-snail striking with rack-hook locking. Bell hammer stop. False plate carried by extended pillars. "Tombstone" segment removed from bottom edge of front and back plate. (Private collection)

4-22. Thomas Dickey thirty-hour brass movement ca. 1820. Anchor recoil escapement. Closed crutch fork. Endless chain with ratchet wheel. Nonrepeat rack-and-snail striking with gathering pallet and rack-pin (dial side) locking. (Private collection)

4-23. George David Eberman eight-day brass movement ca. 1840. Anchor recoil escapement. Closed crutch fork. Grooved drums and tailed ratchet wheel clicks. Nonrepeat rack-and-snail striking with gathering pallet and rack-pin (dial side) locking. Small rectangular segment removed from bottom edge of both plates. (Private collection)

4-24. John Eberman (Jr.) thirty-hour brass movement ca. 1790. Anchor recoil escapement. Closed crutch fork. Endless chain with pulley ratchet wheel having tailless click. Repeat-on-demand rack-and-snail striking with gathering pallet and rack-pin (dial side) locking. (Private collection)

4-25. John Eberman (Jr.) eight-day brass movement ca. 1785. Anchor recoil escapement. Closed crutch fork. Grooved drums with tailed ratchet wheel clicks. Repeat-on-demand rack-and-snail striking with gathering pallet and rack-pin (dial side) locking. Back plate has clipped corners. (Private collection)

4-26. John Eberman (Jr.) eight-day brass movement ca. 1800. Anchor recoil escapement. Grooved drums with tailless ratchet wheel clicks. Open-ended crutch fork. Rack-and-snail striking with modified rack-hook locking. Note that the arbor carrying the rack hook has a detent attached near the back plate which engages the warning wheel pin. Moon wheel and sweep calendar drive wheel missing. (Private collection)

4-27. John Eberman (Jr.) tower clock movement ca. 1785. 53″ high. English pinwheel dead-beat escapement replacing the original Graham dead-beat, part of which is shown in the foreground. Eberman was commissioned to make this clock, for a fee of £550, to replace the first Lancaster County courthouse clock that was destroyed by fire when the first courthouse burned in 1784. This clock served in the second courthouse until 1853, and then in the present courthouse until it was replaced with a Seth Thomas movement in 1898. It is said that son Jacob's career as a clockmaker ended when he lost one of his hands while replacing the hands of this clock around 1796. (Lancaster County Historical Society)

4-28. John Gottlieb Eberman eight-day brass movement ca. 1790. Anchor recoil escapement. Sweep calendar. Grooved drums with tailed ratchet wheel clicks. Closed crutch fork. Repeat-on-demand rack-and-snail striking with gathering pallet and rack-pin locking. Rectangular segment removed from bottom edge of both plates. (Private collection)

4-29. John Joseph Eberman thirty-hour brass movement ca. 1810. Anchor recoil escapement. Tailless ratchet wheel click. Endless chain. Repeat-on-demand rack-and-snail striking with rack-hook locking. Long, narrow segment removed from bottom edge of movement plates. (Private collection)

4-30. John Joseph Eberman eight-day brass movement ca. 1820. Sweep seconds and calendar. Anchor recoil escapement. Closed crutch fork. Smooth drums and tailless ratchet wheel clicks. Rack-and-snail striking with gathering pallet and rack-pin (dial side) locking. Concave ends on vanes of fly. Long, narrow segment removed from bottom edge of movement plates. (Private collection)

4-31. Christian Eby thirty-hour brass movement ca. 1800. Off-center anchor recoil escapement. Open-ended crutch fork. Two-chain pull-up. Ratchet wheels with tailless clicks. Pendulum suspension bridge with elevated suspension post. Repeat-on-demand rack-and-snail striking with rack-hook locking. Pillars pinned to both plates. Front plate has "ear" on lower left to carry dial foot. (Hershey Museum of American Life)

4-32. Jacob Eby eight-day brass movement ca. 1820. Sweep seconds and calendar. Dead-beat pinwheel escapement. Open-ended crutch fork. Smooth drums with tailed ratchet wheel clicks. Repeat-on-demand rack-and-snail striking with rack-hook locking. Narrow bell hammer. Arched segment removed from bottom edge of both movement plates. Front and back plates plugged indicating an earlier layout for a nonsweep movement. (Hershey Museum of American Life)

4-33. John Erb eight-day brass movement ca. 1835. Sweep seconds and calendar. Anchor recoil escapement. Closed crutch fork. Smooth drums with tailed ratchet wheel clicks. Nonrepeat rack-and-snail striking with gathering pallet and rack-pin (movement side) locking. (Private collection)

4-34. John Esterle eight-day brass movement ca. 1816. Anchor recoil escapement. Closed crutch fork. Sweep seconds and calendar. Grooved drums with tailed ratchet wheel clicks. Nonrepeat rack-and-snail striking with gathering pallet and rack-pin (dial side) locking. (Private collection)

4-35. John Esterle eight-day brass movement ca. 1830. Anchor recoil escapement. Open-ended crutch fork. Grooved drums with tailed ratchet wheel clicks. Nonrepeat rack-and-snail striking with gathering pallet and rack-pin (movement side) locking. "J H" inside bell. (Private collection)

4-36. George Fisher thirty-hour brass movement ca. 1800. Anchor recoil escapement. Endless chain with ratchet wheel on pulley. Tailless ratchet wheel click. Open-ended crutch fork. Repeat-on-demand rack-and-snail striking with rack-hook locking. Rectangular segments removed from bottom edge of both plates. (Private collection)

4-37. George Fisher eight-day brass movement ca. 1800. Sweep seconds and calendar. Anchor recoil escapement. Open-ended crutch fork. Smooth drums. Repeat-on-demand rack-and-snail striking with rack-hook locking. Clipped fly vanes. Back plate has the following scratch marks: "H. M. Shreiner, April 14th 1880" (probably Henry M. Shreiner, grandson of clockmaker Martin Shreiner) and "Ben Shreiner, March 6th 1856." Arched segments removed from bottom edge of both plates. (Amish Farm and House)

4-38. George Ford eight-day brass movement ca. 1820. Anchor recoil escapement. Closed crutch fork. Grooved drums with tail-less ratchet wheel clicks. Nonrepeat rack-and-snail striking with gathering pallet and rack-pin (movement side) locking. Bell hammer stop. (Private collection)

4-39. Christian Forrer thirty-hour brass movement ca. 1770. Anchor recoil escapement. Closed crutch fork. Endless chain. Ratchet wheel. Rack-and-snail striking with gathering pallet and rack-pin (dial side) locking. Repeat-on-demand lever missing. Setting disc missing from the alarm unit. Plate pillars riveted to the front plate and pinned to the back plate. (Private collection)

4-40. Christian Forrer eight-day brass movement ca. 1770. Anchor recoil escapement. Closed crutch fork. Grooved drums with tailed ratchet wheel clicks. Repeat-on-demand rack-and-snail striking with gathering pallet and rack-pin (dial side) locking. Plate pillars riveted to front plate and pinned to back plate. Sweep seconds. (Virginia Museum of Fine Arts)

4-41. Christian Forrer eight-day quarter repeater brass movement ca. 1770. Adjustable two-piece recoil escapement (see Fig. 4-83). Closed crutch fork. Grooved drums with tailless ratchet wheel clicks. Nonrepeat rack-and-snail striking with gathering pallet and rack-pin (dial side) locking. Pillars riveted to front plate and pinned to back plate. (Private collection)

4-42. View of quarter repeater train of Christian Forrer movement in Fig. 4-41. When a string is pulled, a small mainspring is wound and the train activated sounding the last hour on the larger hour bell and the last quarter-hour on both bells, e.g., at one minute before 4 o'clock, three blows are struck, followed by three single ting-tangs to indicate the last quarter.

4-43. William Fraser eight-day brass movement ca. 1839. Anchor recoil escapement. Closed crutch fork. Grooved drums with tailed ratchet clicks. Nonrepeat rack-and-snail striking with gathering pallet and rack-pin (movement side) locking. Fly vanes have concave ends. (Private collection)

4-44. Jacob Gorgas thirty-hour brass movement ca. 1770. Anchor recoil escapement. Closed crutch fork. Endless chain. Pulley ratchet wheel with tailless click. Repeat-on-demand rack-and-snail striking with rack-hook locking. Alarm movement missing. (Private collection)

4-45. Jacob Gorgas eight-day brass movement ca. 1771. Anchor recoil escapement. Closed crutch fork. Grooved drums. Rack-and-snail striking with rack-hook locking. Bell mounted within bonnet and activated by wire linkage. Pillars extend through front plate to carry the brass dial. (Lancaster County Historical Society)

4-46. Jacob Gorgas (Jr.) thirty-hour brass movement ca. 1795. Anchor recoil escapement. Closed crutch fork. Endless chain with pulley ratchet wheel and tailless click. Iron pendulum cock with thread loop. Repeat-on-demand rack-and-snail striking with rack-hook locking. Beveled fly vanes. Long pivots. Arched segments removed from bottom edge of both plates. Dial attached to iron "ears" mounted on front plate. Note method of attaching movement to the seat board. (Private collection)

4-47. Jacob Gorgas (Jr.) eight-day brass movement ca. 1790. Anchor recoil escapement. Closed crutch fork. Iron pendulum cock with thread loop. Grooved drums with tailed ratchet clicks. Repeat-on-demand rack-and-snail striking with rack-hook locking. Clipped fly vanes. Long pivots. Arched segments removed from bottom edge of both plates. Extended movement pillars carry the brass dial. (Private collection

4-48. Joseph Gorgas eight-day brass movement ca. 1805. Anchor recoil escapement. Sweep seconds and calendar. Closed crutch fork. Wooden pendulum rod. Smooth drums with tailed ratchet wheel clicks. Repeat-on-demand rack-and-snail striking with gathering pallet and rack-pin (movement side) locking. Hammer stop. Back plate shows signs of previous layout, e.g., pillar holes plugged and plate inverted. (Private collection)

4-49. Solomon Gorgas thirty-hour brass movement ca. 1800. Anchor recoil escapement. Closed crutch fork. Cock pendulum suspension. Endless chain. Pulley ratchet wheel with tailless click. Long pivots. Repeat-on-demand rack-and-snail striking with rack-hook locking. Iron pillars extend through the front plate to carry the dial. Back plate consists of two asymmetrical pieces spliced together. (Private collection)

4-50. Solomon Gorgas eight-day brass movement ca. 1800. Anchor recoil escapement. Closed crutch fork. Brass cock pendulum suspension. Grooved drums and tailed ratchet wheel clicks. Long pivots. Repeat-on-demand rack-and-snail striking with rack-hook locking. Narrow bell hammer. Brass plate pillars extend through the front plate to carry the dial. (Private collection)

4-51. Christian Hall eight-day brass movement ca. 1828. Sweep seconds and calendar. Dead-beat escapement. Pallet arbor has built in by-pass to allow clearance of the third wheel (see Fig. 4-84). Grooved drums with tailed ratchet wheel clicks. Closed crutch fork. Rack-and-snail striking with gathering pallet and rack-pin (dial side) locking. Unusual repeat-on-demand lever on the end of the warning detent. Inside surface of front plate marked "Christian Hall in Litiz fecit April 14th 1828" and "Eby & Flory H-3137." (Private collection)

4-52. William Henry Hall eight-day brass timepiece ca. 1868. Rear view. Dead-beat escapement. Maintaining power. Direct drop weight with winding arbor in rear. Grooved drum. Pin crutch. Iron and brass gridiron (compensated) pendulum with regulation possible at both top and bottom. Secondary motion work on back of the movement drives the hands of a small dial (pictured) in another room. (Private collection)

4-53. John Conrad Heinselman thirty-hour brass movement ca. 1800. Sweep calendar. Anchor recoil escapement. Two chain pull-up. Ratchet wheels with tailless clicks. Pendulum suspension bridge with thread loop. Open-ended crutch fork. Repeat-on-demand rack-and-snail striking with rack-hook locking. Narrow bell hammer. (Hershey Museum of American Life)

4-54. John Conrad Heinselman eight-day brass movement ca. 1790. Sweep seconds and calendar. Anchor recoil escapement. Open-ended crutch fork. Smooth drums. Repeat-on-demand rack-and-snail striking with rack-hook locking. Narrow bell hammer. Arched segment removed from bottom edge of both plates. (Private collection)

4-55. Frederick Heisely and Son eight-day brass movement ca. 1810. Anchor recoil escapement. Smooth drums with keyhole for gut insertion. Tailless ratchet wheel clicks. Pin crutch. Repeat-on-demand rack-and-snail striking with rack-hook locking. This movement is a product of Heisely's later Harrisburg period and is included due to the unavailability of a Lancaster period movement. (Private collection)

4-56. John Hoff thirty-hour brass movement ca. 1800. Off-center anchor recoil escapement. Two-rope pull-up. Pulleys with ratchet wheels having tailless clicks. Open-ended crutch fork. Lantern pinions throughout. Wheels have three crossings. Separate repeat-on-demand lever. Rack-and-snail striking with rack-hook locking. Dial feet carried by three iron "ears" mounted on the front plate. (Private collection)

4-57. John Hoff eight-day brass movement ca. 1800. Anchor recoil escapement. Brass pendulum suspension cock. Open-ended crutch fork. Lantern and open-ended pinions. Smooth drums with ratchet wheels and clicks recessed in the great wheels. Great wheels extend below the bottom edge of the movement to run in a recessed area of the seat board. Keyhole gut insertion slot in drums. Pallet arbor off center. Rack-and-snail striking with separate repeat-on-demand lever. Rack-hook locking. Ten-pound strike weight and seven-pound time weight. Sweep calendar. (Private collection)

4-58. John Hoff eight-day quarter-hour chime brass movement, 1810. Sweep seconds and calendar. Anchor recoil escapement. Open-ended crutch fork. Smooth drums with tailless ratchet wheel clicks. Keyhole for gut insertion. Open-ended and lantern pinions. Rack-and-snail striking with rack-hook locking. Dovetailed bell hammer spring. Five-note quarter-hour chime on nest of five bells. Originally purchased in 1810, without case, for £26. (Lancaster County Historical Society)

4-59. John George Hoff thirty-hour brass movement ca. 1770. Anchor recoil escapement. Iron pendulum suspension cock with thread loop. Endless chain with pulley ratchet wheel having tailless click. Three crossings per wheel. Open-ended pinions. Rack-and-snail striking with rack-hook locking. Separate repeat-on-demand lever. Iron pillars extend through front plate to carry the iron dial. (Collection of the Pennsylvania Farm Museum of Landis Valley)

4-60. John George Hoff eight-day brass movement ca. 1785. Anchor recoil escapement. Brass-pendulum suspension cock. Open-ended crutch fork. Smooth drums with keyhole for gut insertion. Tailless ratchet wheel clicks recessed, with ratchet wheel, within solid great wheels. Open-ended and lantern pinions. Rack-and-snail striking with rack-hook locking. Separate repeat-on-demand lever. Dovetailed bell hammer spring. (Private collection)

4-61. John George Hoff eight-day brass movement ca. 1800. Sweep seconds. Anchor recoil escapement. Open-ended crutch fork. Smooth drums with tailed ratchet wheel clicks. Keyhole slot for gut insertion. Open-ended and lantern pinions. Rack-and-snail striking with rack-hook locking. Separate repeat-on-demand lever. Dovetailed bell hammer spring. (Private collection)

4-62. Probable John George Hoff thirty-two-day brass movement ca. 1800. Anchor recoil escapement. Smooth drums with keyhole for gut insertion. Recessed ratchet wheels with tailless ratchet clicks. Open-ended crutch fork. Brass pendulum suspension cock. Lantern pinions throughout. Rack-and-snail striking with rack-hook locking. Separate repeat-on-demand lever missing. Five pillars separate the plates. It incorporates nearly all the Hoff features. The bell hammer spring resembles that found on the earlier Hoff thirty-hour movements. (Private collection)

4-63. John George Hoff eight-day brass musical movement ca. 1810. Anchor recoil escapement. Brass pendulum suspension cock. Open-ended crutch fork. Lantern pinions. Rack-and-snail striking with rack-hook locking. Separate repeat-on-demand lever. Manual select ten-tune brass music cylinder with nest of fourteen bells. Automatic night music train silence lever. (Private collection)

4-64. John George Hoff eight-day brass musical movement. Left rear quarter view of foregoing movement. Quarter-hour strike bell nested inside hour bell mounted on back plate. Note solid great wheel with recessed ratchet wheel and click. Music repertoire consists of: "March, Rigadoon, Menuetto, Meine Herr, Allemenschen, Menuetto, Marsch, Pollones, Nun Last uns Gott, and Trumpeter." (Private collection)

4-65. John George Hoff, Jr. thirty-hour brass movement ca. 1810. Anchor recoil escapement. Two-chain pull-up. Ratchet wheels with tailless clicks. Open-ended crutch fork. Lantern pinions. Rack-and-snail striking with rack-hook locking. Separate repeat-on-demand lever. "Ear" on lower right corner of the front plate carries a dial foot. (Private collection)

4-66. John George Hoff, Jr. eight-day brass movement ca. 1815. Sweep seconds and calendar. Anchor recoil escapement. Smooth drums with keyhole for gut insertion. Tailless ratchet wheel clicks. Seatboard cut to permit clearance of great wheels that extend below plate bottom edges. Open-ended crutch fork. Open-ended and lantern pinions. Repeat-on-demand rack-and-snail striking with rack-hook locking. Dovetailed bell hammer spring. (Private collection)

4-67. Jacob Kammerer thirty-hour brass movement ca. 1800. Anchor recoil escapement. Two chain pull-up with ratchet wheels. Repeat-on-demand rack-and-snail striking with rack-hook locking. Note unusual bell hammer tail. Front and back plates have arched segments removed from their bottom edge. (Collection of the Pennsylvania Museum of Landis Valley)

4-68. John Leinbach thirty-hour brass movement ca. 1795. Anchor recoil escapement. Open-ended crutch fork. Endless chain with pulley ratchet wheel and tailless click. Repeat-on-demand rack-and-snail striking with rack-hook locking. Bell marked "G. Ainsworth Warrn." (Warrington, England ca. 1810). Pillars screwed to the back plate. An intermediate wheel drives a sweep seconds hand (see Fig. 4-85). (Private collection)

4-69. Henry Lewis Montandon eight-day brass movement ca. 1800. Anchor recoil escapement. Altered crutch fork. Grooved drums with tailless ratchet wheel clicks. Nonrepeat rack-and-snail striking with gathering pallet and rack-pin (dial side) locking. Fly vanes have been widened. (Collection of Richard W. Case)

4-70. Henry Ober (Jr.) eight-day brass movement ca. 1830. Anchor recoil escapement. Closed crutch fork. Grooved drums with tailed ratchet wheel clicks. Repeat-on-demand rack-and-snail striking with rack-hook locking. Narrow bell hammer. Arched segment removed from bottom edge of movement plates. (Private collection)

4-71. Samuel Quest eight-day brass movement ca. 1820. Anchor recoil escapement. Smooth drums with tailless ratchet clicks. Open-ended crutch fork. Nonrepeat rack-and-snail striking with gathering pallet and rack-pin (dial side) locking. Sweep seconds and calendar. Arched segments removed from bottom edge of front and back plates, top corners have cyma reversa pieces removed. (Private collection)

4-72. Martin Shreiner thirty-hour brass movement ca. 1800. Anchor recoil escapement. Closed crutch fork. Endless chain with ratchet wheel on pulley and tailless click. Pear-shaped weight similar to a counterpoise of a steelyard scale. Repeat-on-demand rack-and-snail striking with gathering pallet and rack-pin locking. Trapezoidal section removed from bottom edge of both plates. (Collection of the Pennsylvania Farm Museum of Landis Valley)

4-73. Martin Shreiner eight-day brass movement ca. 1798. Sweep seconds and calendar. Anchor recoil escapement. Closed crutch fork. Grooved drums. Rack-and-snail striking with gathering pallet and rack-pin locking. Graceful levers. Trapezoidal section removed from bottom edge of both movement plates. (The James Buchanan Foundation for the Preservation of Wheatland)

4-74. Martin Shreiner eight-day brass musical movement ca. 1810. Anchor recoil escapement. Closed crutch fork. Grooved drums with tailed ratchet wheel clicks. Repeat-on-demand rack-and-snail striking with gathering pallet and rack-pin (dial side) locking. Hour bell marked "G. Ainsworth Warrn." (Warrington, England ca. 1810). Self Changing seven-tune brass music cylinder with nest of seven bells, each with two hammers. Single let-off pin for separate music and strike levers. Trapezoidal segments cut from the bottom edge of both plates. Sweep seconds and calendar. (Private collection)

4-75. Martin (Jr.) and Philip Shreiner eight-day brass movement ca. 1835. Sweep seconds and calendar. Dead-beat escapement with the anchor straddling the third arbor of the time train. Closed crutch fork. Grooved drums with tailed ratchet wheel clicks. Nonrepeat rack-and-snail striking with gathering pallet and rack-pin (movement side) locking. (Private collection)

4-76 George Smith eight-day brass movement ca. 1815. Anchor recoil escapement. Open-ended crutch fork. Grooved drums with tailless ratchet wheel clicks. Nonrepeat rack-and-snail striking with gathering pallet and rack-pin (dial side) locking. Bell marked "John Wilbank Caster Phila." (Philadelphia, Pa. 1813 to 1842). (Private collection)

4-77. Samuel C. Stauffer eight-day brass movement ca. 1795. Sweep seconds. Dead-beat escapement. Open-ended crutch fork. Smooth drums with tailed ratchet wheel clicks. Repeat-on-demand rack-and-snail striking with rack-hook locking. Arched segments removed from bottom edge of movement plates. Pillars pinned front and back, and extend to carry the brass dial. (Hershey Museum of American Life)

4-78. Rudy Stoner thirty-hour brass movement ca. 1755. Anchor recoil escapement. Closed crutch fork. Endless chain with iron hoop spring and great wheel spoke ratcheting. Repeat-on-demand rack-and-snail striking with gathering pallet and rack-pin (dial side) locking. Brass pillars extend through the front plate to carry the brass dial. (Private collection)

4-79. Gustavus Stoy eight-day brass movement ca. 1800. Anchor recoil escapement. Closed crutch fork. Grooved drums with tailless ratchet wheel clicks. Repeat-on-demand rack-and-snail striking with rack-hook locking. All wheels except the pin wheel and escape wheel are mounted on the pinions instead of collets. Sweep seconds and calendar. This movement was made during Stoy's earlier Lebanon, Pennsylvania period, and is included due to the unavailability of a Lancaster movement. (Private collection)

4-80 Abel Witmer thirty-hour brass movement ca. 1795. Anchor recoil escapement. Closed crutch fork. Endless chain with ratchet wheel on pulley. Tailless ratchet click. Repeat-on-demand rack-and-snail striking with rack-hook locking. Spring and pin override on index end of rack tail. Long pivots. Iron pendulum suspension cock. (Private collection)

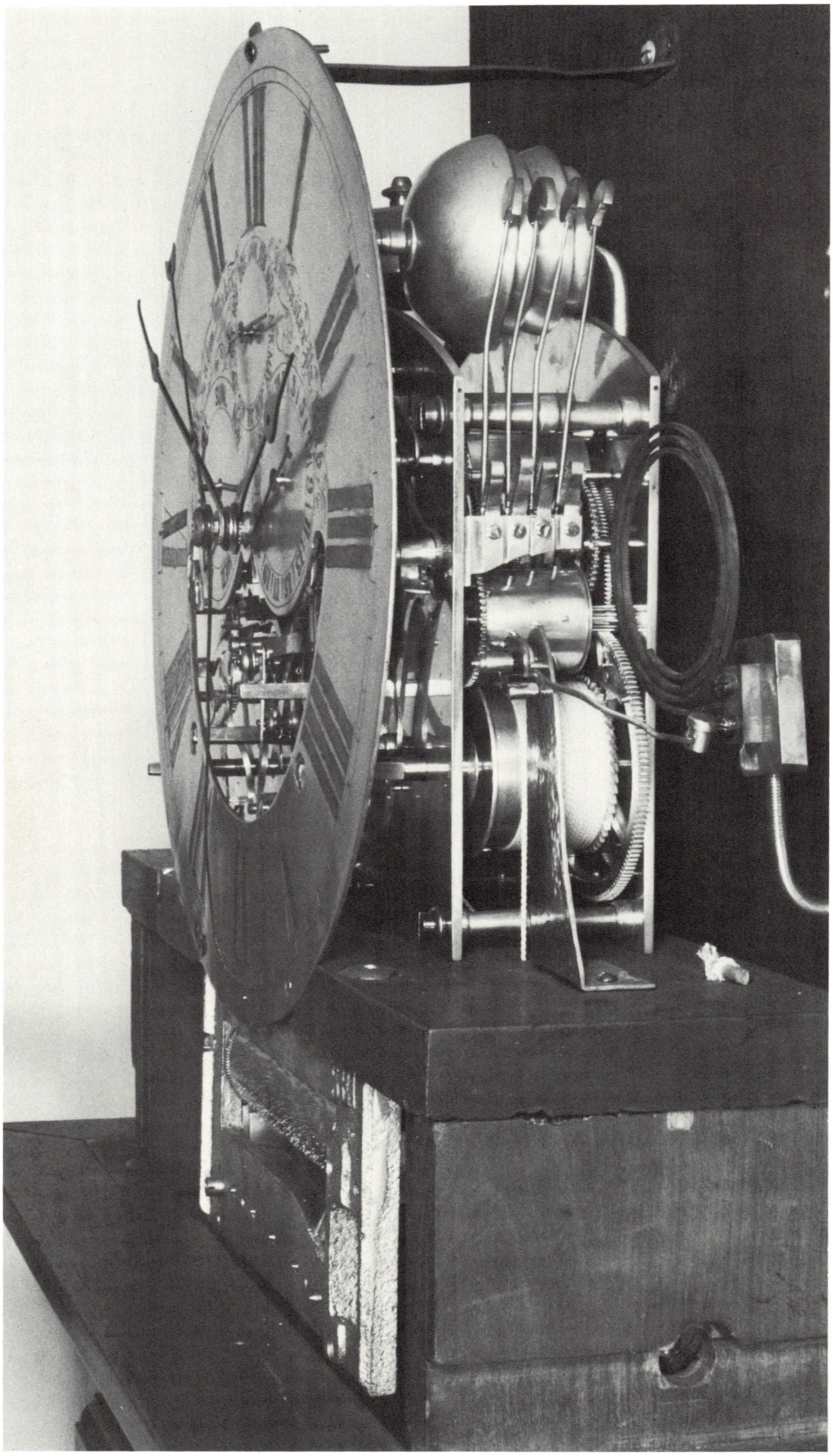

4-81. Godfried M. Zahm eight-day three-train brass movement, 1876. Dead-beat escapement. Smooth drums. Rack-and-snail striking with gathering pallet and rack-pin locking. Auxiliary "last hour struck" and music box movements. Sequenced to chime the quarter-hour on a nest of four bells and coiled gong, to strike the hour past on a separate gong and the present hour on a bell, and to play one of four tunes on a music box (self shifting). (Private collection)

4-82. Unsigned eight-day brass movement. Anchor recoil escapement. Smooth drums with keyhole for gut insertion. Open-ended crutch fork. Rack-and-snail striking with rack-hook locking. Separate repeat-on demand lever. Dovetailed bell hammer spring. This movement has all the Hoff traits except that it has drawn pinions instead of lantern or open-ended wire pinions. It may be the work of a Hoff or a Hoff apprentice. (Lancaster Mennonite Conference Historical Society)

4-83. Christian Forrer movement (Fig. 4-41) viewed from top, front. At the right is the main arbor assembly of the repeat mechanism, containing (front to back) the great wheel and ratchet, the count wheel, and the mainspring housing. In the center and left is the rare, double-pallet recoil escapement invented by Chevalier de Béthune and applied by M. Thiout of Paris in 1727.[1] Note that the teeth of the escape wheel are reversed in comparison to those of the conventional anchor recoil design. The exit pallet is to the right and is mounted on a conventional pallet arbor and crutch assembly; the screw permits adjustment of the escapement. The counter-weighted entrance pallet is to the left and is mounted on a separate, free swinging arbor. (Private collection)

1. B. Thorpe, *Watch and Clock Escapements* (Philadelphia: Keystone Press, 1904)

4-84. Detail of Christian Hall movement (Fig. 4-51) showing the by-pass in the escape arbor (mirrored in the back plate) to allow clearance of the third wheel of the time train. Another Hall movement examined did not have this feature. (Private collection)

4-85. Detail of John Leinbach thirty-hour movement (Fig. 4-68) showing the gearing allowing for a clockwise sweep seconds hand. (Private collection)

4-86. Eight-day brass shelf clock movement ca. 1820 behind a dial marked "Sam'l S. Grosch/Marietta." Exterior anchor recoil escapement (partially obscured from view by the large-toothed calendar wheel). Compounded weights. Sweep seconds and calendar. Repeat-on-demand rack-and-snail striking with gathering pallet and rack-pin (dial side) locking. (Private collection)

INDEX OF CLOCKS

The following tabular index will enable you to "re-assemble" many of the clocks. It should be noted that not all clocks are represented in all three sections, i.e., case, dial, and movement.

Name	Case Fig.	Dial Fig.	Movement Fig.
Hoff, John George			
30-hour	2-3	3-43	4-59
8-day	2-4	3-45	4-60
8-day		3-47	
8-day	2-2	3-44	
8-day, sweep seconds		3-48	4-61
8-day, sweep seconds	2-11	3-46	
8-day, musical	2-32	3-49	4-63, 64
32-day			4-62
Hoff, John George, Jr.			
30-hour		3-50	4-65
8-day, sweep seconds		3-51	4-66
Kammerer, Jacob			
30-hour			4-67
Leinbach, John			
30-hour, sweep seconds	2-26	3-52	4-68, 85
Montandon, Henry Lewis			
8-day		3-53	4-69
Ober, Henry (Jr.)			
8-day	2-52	3-54	4-70
Quest, Samuel			
8-day, sweep seconds	2-48	3-55	4-71
8-day, sweep seconds		3-56	
Shreiner, Martin			
30-hour	2-6	3-57	4-72
8-day, sweep seconds	2-12		
8-day, sweep seconds			4-73
8-day, musical	2-34	3-58	4-74
Shreiner, Martin (Jr.) & Philip			
8-day, sweep seconds	2-49	3-59	4-75
Smith, George			
8-day	2-43	3-60	4-76
Stauffer, Samuel C.			
8-day, sweep seconds	2-18	3-61	4-77
Stoner, Rudy			
30-hour	2-1		4-78
30-hour	2-9	3-62	
Stoy, Gustavus			
8-day, sweep seconds			4-79
Witwer, Abel			
30-hour	2-28	3-63	4-80
Zahm, Godfried M.			
8-day, calendar	2-54	3-64, 65	4-81
Unsigned			
8-day	2-22		4-82

GENERAL INDEX

Page numbers for illustrations are in italics.